City of Steel

City of Steel

How Pittsburgh Became the World's Steelmaking Capital During the Carnegie Era

Ken Kobus

ROWMAN & LITTLEFIELD
Lanham • Boulder • New York • London

Published by Rowman & Littlefield
A wholly owned subsidiary of The Rowman & Littlefield Publishing Group, Inc.
4501 Forbes Boulevard, Suite 200, Lanham, Maryland 20706
www.rowman.com

Unit A, Whitacre Mews, 26-34 Stannary Street, London SE11 4AB

British Library Cataloguing in Publication Information Available

Library of Congress Cataloging-in-Publication Data

Kobus, Ken.
City of steel : how Pittsburgh became the world's steelmaking capital during the Carnegie era / Kenneth J. Kobus.
pages cm
Includes bibliographical references and index.
ISBN 978-1-4422-3134-4 (cloth : alk. paper) -- ISBN 978-1-4422-3135-1 (electronic)
1. Steel industry and trade--Pennsylvania--Pittsburgh--History--20th century.
2. Pittsburgh (Pa.)--History--20th century. I. Title.
HD9518.P57K63 2015
338.4'7669142097488609034--dc23

2014049895

∞™ The paper used in this publication meets the minimum requirements of American National Standard for Information Sciences—Permanence of Paper for Printed Library Materials, ANSI/NISO Z39.48-1992.

Printed in the United States of America

This book is dedicated to my wife, Linda, who spent many long, boring hours with me in libraries and archives while I was pondering the mysteries of steel, and for listening to and reading my work when I know that she really wished to be doing almost anything else.

Contents

Foreword

The author has produced a book that will be required reading for all students of the steel industry. Beginning with a detailed history of the development of early steelmaking up through the time of Andrew Carnegie, some startling insights are revealed. Why was the original American steelmaking centered in Pittsburgh? Was Andrew Carnegie really the one-sided man history has drawn him to be? Why was Carnegie so successful in attracting and motivating talented subordinates? These and other characteristics of the man—not previously identified—are studied in some detail. Andrew Carnegie emerges as a much more enlightened and warmer figure than previous biographers have described. The book is meticulously researched and annotated. This author should be applauded for challenging some long-held mythology surrounding Andrew Carnegie.

T. C. Graham

Mr. Graham is the former president and CEO of Jones and Laughlin Steel, president of United States Steel Corporation or USX Corporation, chairman and CEO of Washington Steel Corporation, chairman and CEO of AK Steel, and president and CEO of Armco Steel Company, LP.

Acknowledgments

I thank the following persons for their contributions: first and foremost I thank Dr. Kenneth Warren, Emeritus Fellow, University of Oxford, Jesus College, for many years of friendship, discussion, guidance, and encouragement; Tom Graham, former president of United States Steel Corporation, for believing in and endorsing this project; Bill Roemer, for his much-needed and welcomed assistance in editing the text; Ron Baraff, director of Museum Collections and Archives; Tiffani Emig, curator; and August Carlino, president and CEO of the Steel Industry Heritage Corporation's Rivers of Steel National Heritage Area in Homestead, Pennsylvania, for reviewing my work, assisting in my searches, and being a sounding board; Lawrence John, MD, my personal physician, a friend, and a constant fountain of optimism; Jim Saunders, my boss, who taught me much about the interworkings of the mill, Joel Tarr, Richard S. Caliguiri Professor of History and Policy, Heinz College at Carnegie Mellon University; David Hounshell, David M. Roderick Professor of Technology and Social Change at Carnegie Mellon University, and Edward K. (Ted) Mueller, Department of History of the University of Pittsburgh, for being advocates and supporters of this and other historical research projects of mine; Rod Sellers, Southeast Chicago Historical Society; Gil Pietrzak, librarian, Pennsylvania Department, Carnegie Library of Pittsburgh; Adam Ryan, Carnegie Museum of Art, Pittsburgh; Miriam Meislik, University of Pittsburgh Archives Service Center; Sierra Green and Alexis Macklin, Senator John Heinz History Center; Lynsey Sczechowicz, Hagley Museum and Library; many helpful reference librarians at Carnegie Library of Pittsburgh, Main Library, as well as reference librarians at the University of Pittsburgh, Hillman Library for assisting me with the numerous reference, book, and photo searches required by this project; and finally to Jed Lyons of Rowman & Littlefield, who provided me the opportunity to publish this work.

Introduction

The dominant position that Pittsburgh held in the world of heavy industry is certainly well documented and, as far as the production of iron and steel is concerned, the leadership role that it held over the entire world is unquestionable. Legions of people identify this city with the steel industry, and those associations became so intimately interwoven that the words *Pittsburgh* and *steel* became synonymous. This idea still prevails, despite the fact that in today's world the region's manufacturing output is miniscule when compared with the past or to that of other areas of the United States, and especially so when measured against the rest of our planet. Still, while Pittsburgh would have been an important steelmaking district, I suggest to you that it should *not* have become the steelmaking capital of the world.

For instance, Eber Ward rolled the first steel rail in the United States, which was made in Chicago in 1865. The steel for that rail was produced in Wyandotte, Michigan, where the nation's first Bessemer steel was made under the direction of William Durfee. In 1870, a visiting German concluded that St. Louis offered the soundest foundation for the manufacture of iron and Bessemer steel.[1] During 1889, for example, the newly formed "Illinois Steel Company was believed to have a larger output than any other steel company in the world, although it employed fewer men than the Krupp works at Essen."[2] The Illinois Steel Company's plants were located in and around the Chicago area. What, then, led Pittsburgh down the path to its commanding position over the world's steel industry? Since the reasons for Pittsburgh's success have never been fully presented by any historian, the answer requires that one first understand the methods involved in manufacturing steel, which has become one of our modern world's most basic requirements. Late nineteenth-century America was on the cusp of change from an agrarian to an industrial economy. The materialization of this county's steel industry was a driving force behind that transition.

During the middle third of the nineteenth century in the Pittsburgh region, a large industrial base was developed through the manufacture of wrought iron followed by the making of crucible steel. Soon these new materials were used in the fabrication of many useful goods. They became widely distributed throughout America's rapidly emerging Western Reserve on this side of the Allegheny Escarpment, aided by the extensive river network flowing from the Pittsburgh gateway in that direction. The mountains were a major impediment to shipping these same goods to the more civilized society in the east, requiring either a long walk or a slow ride on your choice of the Erie or Pennsylvania canals, or perhaps a trip on the fledgling Pennsylvania Railroad.

The demand for mass-produced steel, now that tons instead of pounds could be made in minutes instead of hours, was increasing in order to fulfill the needs of westward railroad expansion. Eastern and Midwestern Bessemer plants, operating for nearly a decade beginning in the 1860s, were accommodating the railroads in the conversion from wrought iron to steel rails. Pittsburgh did not benefit from this increased demand because fast steelmaking plants had not yet been constructed there. Coal was abundant in the Pittsburgh region and is an important aspect of this history, but other industrial cities also had ready access to coal.

Although numbers of different businesses were responsible for the birth of the metals industries in the Pittsburgh area, the crux of the tale primarily revolves around the enterprises of one man, Andrew Carnegie, and his Carnegie Steel Company. The links between the various synergies developed by these firms has largely gone unexplored and so for the most part is not well known. These connections for success are fairly obvious but are sometimes hidden in plain sight, being often obscured beneath the harsh light of modern thinking. This is somewhat akin to the paradox of the chicken and the egg: Who were these pioneers? How do you define that term? Why?

There are a number of distinct reasons for Pittsburgh's rise to the top of the steel industry that will be explored. Some apparently disassociated items can be seen to converge in compatible modes that formed these synergies. Technology, invention, innovation, automation, management style, conservation and efficient use of materials, insight, planning, science and reasoning were all used to build successful enterprises. In a few instances, overcoming resistance to change took locally failed factories and transformed them into successful enterprises with the same employees.

This is an account that has needed to be written concerning a revolutionary period in American history. As a third-generation steelworker, who has a deep and intense interest in the processes of steelmaking and the history of

Pittsburgh, I felt an obligation to share my interest and experiences, not by describing the evolution in the technology using mathematical or scientific formulas, but by presenting easily understood word descriptions for those uninitiated in the birthing of steel. I have found that it is also necessary to recount the sometimes difficult and often brutal conditions that were frequently endured by some of those workers in the process. As the links between many of these technical, innovative, and economic associations have never been explored, you will see how the dots have been connected to answer the questions of how and why Pittsburgh *did* become the Steel Capital of the World.

Part I

PITTSBURGH ENTERS THE METALS INDUSTRY

Chapter One

Wrought Iron and Puddling

While wrought iron is not steel, it is an important precursor material to mass-produced steel and was used for making machinery, equipment, and rails. Because America was in the early stages of transition to an industrial economy, rails were important. It is also beneficial to recognize the contribution that wrought iron manufacture made and the part it played in the development of the modern steel industry.

Manufacturing wrought iron was accomplished in the past through the reduction of iron ore by charcoal.[1] In 1784 Henry Cort developed a process in England whereby wrought iron was made from cast iron. The manufacture of the wrought product was undertaken in a smallish rectangular box, officially known as a reverberatory furnace. It had a hollowed-out depression in the center of its sand floor. The benefit was that the process was accomplished through indirect heating by burning a mineral fuel instead of using more expensive charcoal.[2] (For a firsthand description of the Cort process, see John Percy, *A Treatise on Metallurgy: Iron and Steel.*)

The reverberatory furnace was frequently described as an *air furnace*, because the material being heated did not contact the fuel, only the flame. The furnace was simply a refractory-lined enclosure, roughly a rectangular chamber, whose floor or *hearth* was about waist high above the normal grade. Material losses were about 30 percent higher using Cort's method due to slagging of the iron.[3] An important aspect of this new process was that it enabled the production of at least two-and-one-half times more than was previously possible.[4] Because the hole or depression in the hearth became filled with the molten metal, usually called the *bath*, it reminded workers of a puddle of iron, and it became known as a *puddling furnace*. Baldwin Roger introduced a cast iron hearth in 1818,[5] and Joseph Hall improved the process around 1830 by using mill scale (iron oxide) on the cast-iron-lined hearth, which made it chemically a base. This reduced losses to about 10 percent. At molten metal temperatures, sand exhibits the same chemical properties as an acid, whereas the scale and cast plates used for the furnace floor by Hall exhibited the chemical properties of a base. With Hall's improvement, large volumes of slag evolved from working the iron, so it became known as *wet puddling*. Metallurgists of the time referred to Cort's as the *dry puddling* process.[6]

Wrought iron manufacturing began in the Pittsburgh area about 1817 through Isaac Meason's experiments at Plumsock, Pennsylvania (it was formerly known as Middleton or Middletown, but at the present time there is no political entity near that location identified with any of those given names).[7] The town was situated on Redstone Creek midway between Connellsville and Brownsville at a point east of present-day Waltersburg.[8] In 1826 there were four mills operating in the area.[9] By the 1840s Pittsburgh had become the wrought iron manufacturing center of the United States. Like steel, wrought iron was usually made from *cast* or *pig* iron by reducing the amount of carbon, phosphorous, sulfur, and other detrimental materials that made the iron very brittle. In the case of wrought iron, essentially all of the carbon was removed. This created a ductile substance that could be rolled or hammered into useful shapes that had high strength and resistance to breaking.

The puddling furnace (see figure 1.1) had a grate at one end on which coal was usually burned. In Pittsburgh, bituminous coal was usually used, but some furnaces used natural or produced gas, while other regions used anthracite coal or even coke. The flame from the fire drafted over a low wall and was then deflected or reverberated downward by the roof (hence the official designation of reverberatory furnace) onto the hearth where the product was made. The flame could be adjusted using a damper on the chimney at the opposite end of the furnace. Working the damper, the *puddler* (the name given the operator) could change the furnace enclosure from an oxidizing to a reducing atmosphere through increasing or decreasing the amount of air entering it. A smoky environment inside the hearth would have little oxygen, consequently resulting in a chemically reducing surrounding. Creating a clear, bright flame through opening the damper and increasing the draft increased the flow of air, creating an oxidizing atmosphere. Adjusting the damper would also slightly raise or lower the temperature.

Making wrought iron followed five distinct stages: melting, clearing, boiling, balling, and drawing.[10] Sometimes puddling was called *pig boiling*,[11] although there is a fine distinction, making that name inaccurate.[12] The puddler had his helper load the furnace with about five hundred to six hundred pounds (typical, although furnaces ranged in capacity from about 300 to 1,500 pounds). The cast iron usually contained between 3 or 4 percent carbon. About fifty pounds of mill scale was added (small flecks of oxidized iron that

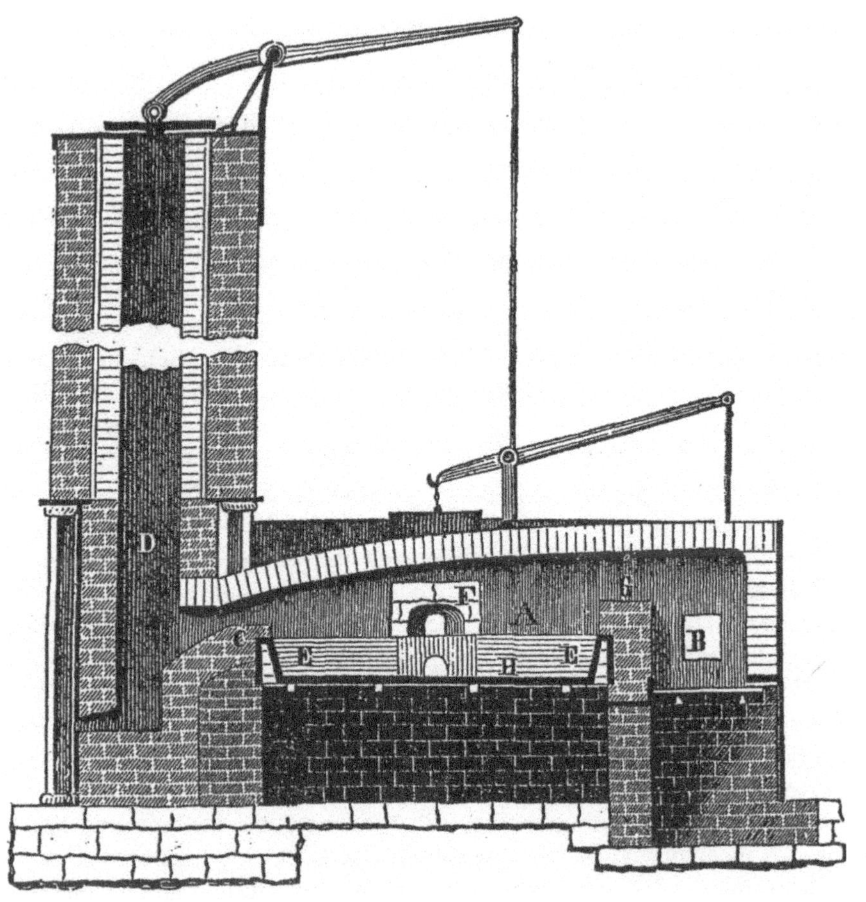

Figure 1.1. A diagram of a reverberatory or puddling furnace. A is the furnace chamber, B is the grate area where the fuel was burned, H is the hearth, F is the working door, and D is the chimney with a damper on top. Note how the roof slopes downward to deflect or reverberate the flame from the combustion chamber (far right) onto the hearth after first surmounting the division wall.

From Frederick Overman, Metallurgy, 1855.

would fall off of the metal as it was rolled in the mill). The helper shoveled about five hundred pounds of coal onto the grate. As the flame from the coal fire surmounted a low wall on the way to the stack, it was directed onto the iron in the hearth by a downward-sloping roof, where it impinged upon the newly charged iron that became molten after about half an hour.[13, 14] The puddler worked the iron through a stomach-high opening about twenty inches square in the side of the furnace, sometimes called the *working door.*[15] While the iron was being melted and stirred or *rabbled*, the puddler watched to see that both silicon and manganese was being removed. James Davis, a former puddler, provided this description of the work: "I was stirring the charge with a long iron rabble that weighed some twenty-five pounds. Strap . . . that weight to your arm and then do calisthenics ten hours in a room so hot it melts your eyebrows and you will know what it is like to be a puddler."[16] When the iron melted, the operator reported that the iron was *cleared*.

Technically, the flame was oxidizing the silicon in the iron (burning it, through the high temperature and excess air of the flame) as it was simultaneously oxidizing some of the iron itself during the melting stage. Additionally, the charge became oxidized through contact with the iron-oxide lining of the furnace, and this formed a basic slag during melting. Slag formation was a mechanism through which the silicon and manganese were eliminated. This typically occurred in the last eight to ten minutes of melting. When the silicon was removed or *cleared*, it "markedly changed the character of the iron."[17]

During the *clearing stage*, a "proper" amount of mill scale or good-quality iron ore was added to the mix to make more basic slag. Adding the scale, in addition to adjusting (closing) the damper and sometimes by throwing a bit of water into the furnace, would lower the temperature of the bath.[18] The lower temperature favored the removal of phosphorus and sulfur "ahead of" or before the consumption of the carbon. This part of the cycle was referred to as the *low boil*. As he stirred, "the puddler watched the color of the metal for a subtle change from a reddish to a bluish hue, which showed that as much phosphorus as could be removed had been removed."[19] As soon as the temperature was increased again (by opening the damper and waiting for the flame to act), the carbon began to be oxidized more rapidly (because of the elevated temperature the carbon combined with the oxygen of the flame and/or oxygen from the mill scale or iron ore, both of which are iron oxides). As a result, carbon monoxide gas began to evolve in ever-larger quantities. Where it broke through the layer of slag, the carbon monoxide caused little blue flames called *puddler's candles* to form in increasing numbers.[20] The gasses being generated inside caused the mass to swell, entering a stage called the *high boil*. More slag flowed from the furnace. This phase of the operation lasted twenty to twenty-five minutes. The puddler continued to rabble the mixture. Davis wrote,

> More than an eighth and sometimes a quarter of the weight of the pig-iron flows off in the slag. Meanwhile I have got the job of my life on my hands. I must stir the boiling mess with all the strength in my body. For now is my chance to defeat nature and wring from the loosening grip of her hand the pure iron she never intended to give us.[21]

As the metal became purer, the furnace temperature (at its maximum about 2,600° F[22] or 1,427° C) was not sufficient to keep it molten. Bits of wrought iron began to precipitate out of the bath. Some nearly pure iron would poke through the slag, while some would form a pasty mass on the bottom of the furnace. At this point it was said that the metal *comes to nature*.[23] The puddler had to work the heat in such a manner that the material poking through the slag did not become reoxidized as well as the material on the bottom did not chill and attach itself to the hearth. He had to push exposed material into the bath and pull the other off of the bottom, making sure that the phosphorus in the slag did not return (revert) to the metal.

> Little spikes of pure iron like frost spars glow white-hot and stick out of the churning slag. These must be stirred under at once; the long stream of the flame from the grate plays over the puddle, and the pure iron if lapped by these flames would be oxidized—burned up. Pasty masses of iron form at the bottom of the puddle. There they would stick and become chilled if they were not constantly stirred. The whole charge must be mixed as it steadily thickens so that it will be uniform throughout. I am like some frantic baker in the inferno kneading a batch of iron for the devil's breakfast.

This is again from Davis' account.[24] As larger amounts of wrought iron were collected, the puddler had to divide the material into several *balls*.

> The "balling" of this sponge into three loaves is a task that occupies from ten to fifteen minutes. The particles of iron glowing in this spongy mass are partly welded together; they are sticky and stringy and as the cooling continues they are rolled up into wads like popcorn balls.[25]

Depending on the size of the furnace, these balls could range in size from 125 to 300 pounds, but just under two hundred pounds was more typical.

The entire operation took one hour and ten minutes to an hour and forty-five minutes to complete.[26] Average production for a furnace was about 2,800 pounds in four to five heats per day. The *puddle balls*, still dripping with slag, were extracted from the furnace using tongs, and they were placed in a mechanical device called a *squeezer*, where much of the excess slag from the ball would be removed. It was formed into a shape that could easily be inserted between the rolls of a mill. There it would be rolled into a bar, usually called a *muck bar* or *puddle bar*. During the rolling process, more slag was pressed from the wrought iron, giving it a fibrous texture of iron, interspersed with strings of slag. At facilities where there was no squeezer, the balls would be hammered into a bloom in a forge, or *shingled*, and then rolled.

In the early days of wrought iron manufacturing, when the double refined ball was removed from a *Catalan forge*, it was called a "blume," because it looked like a flower. Today in English, an intermediate product rolled from an ingot is called a *bloom*, derived from the German *blume* for flower.[27]

Fuel requirements for this process were about one ton of coal for each ton of wrought iron produced.[28]

A puddler's skill could not be learned easily. His job required a great deal of strength and stamina. More than that, puddling required the operator to:

- create change in the furnace atmosphere from oxidizing to reducing and vice versa by adjusting a damper.
- discern the nuances of color indicative of a temperature and/or a chemical change; that is, that the iron had cleared or that the maximum amount of phosphorus had been removed.
- multitask to prevent the refined iron from reoxidizing.
- sustain furnace reactions.

The puddler had a certain degree of power. His skill was often passed from father to son, or he would take someone of his choosing as an apprentice. Frequently, the company paid the puddler, and he in turn paid his helper. This gave him even more power and leverage with his employer. Based on the number of heats produced, within reason the puddler set the production rate for each day. Even the owner of the company could not compel him to produce a greater output. He could demand more, but it was unlikely to happen. Furthermore, the proprietor probably did not possess the expertise to make it happen on his own.

The amount of control wielded by the puddler, the small output per furnace, the immense quantity of wrought iron needed by manufacturers in the Pittsburgh area and elsewhere, as well as the great numbers of puddlers required gave rise to trade unionism in the district. Skilled craftsmen banded together to form an organization called the United Sons of Vulcan. Conflict between capital and labor would become a recurring theme in the history of the Pittsburgh region as the balance of power ebbed and flowed between these two competing factors.

The mysterious "happenings" that occurred within the confines of the puddling furnace can be easily explained through the use of a device called a *phase diagram* (see figure 1.2). Since pig iron, wrought iron, and steel are all mixtures of iron and carbon, albeit at different percentages, one can interpret the behavior of the material in the furnace by using such a diagram. The diagram displays the physical nature (whether solid, liquid, or a combination of both) of iron versus the differing percentage of carbon contained within the mixture. In other words, the diagram showed the effect that furnace temperature had on the physical state of iron, containing varying proportions of carbon. Once heated, the furnace was operated at a fairly constant temperature throughout the process. One can merely follow the horizontal line representing the normal maximum operating temperature (about 2,600°F

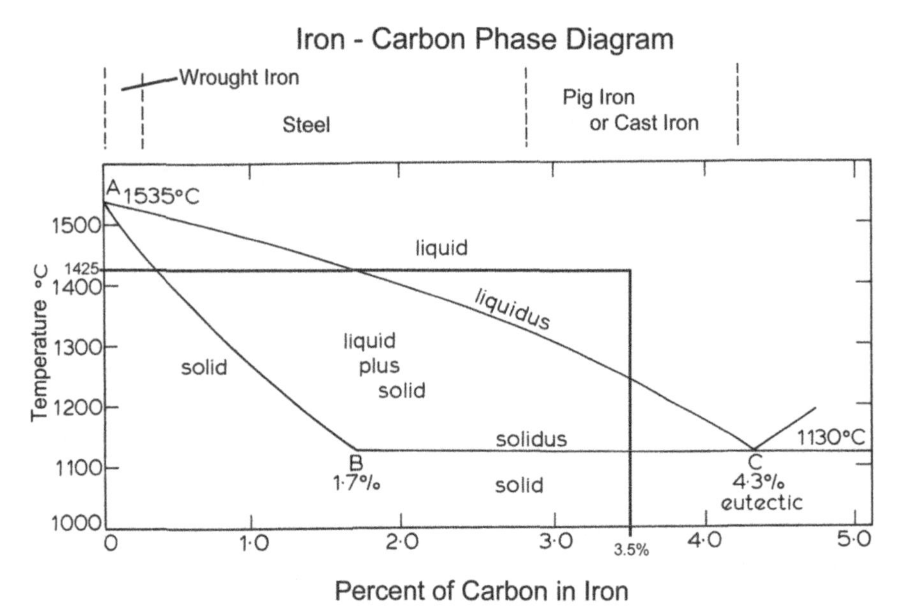

Figure 1.2. A metallurgical "Phase Diagram" of a mixture of iron and carbon. The heavy dark vertical and horizontal lines superimposed on the graph are indicative of both the composition of the metal charge and the temperature of the furnace.
By the author, based on one from Barraclough, Crucible Steel.

or 1,425°C) to its intersection with a vertical line representing the percent of carbon contained within the charge of pig iron.

Figure 1.2 shows that where the vertical line indicating 3.5 percent carbon content (a typical value) intersects with a horizontal line marked *solidus* (at about 1,130°C), the iron begins to melt. The intersection with the sloping (*liquidus*) line at about 1,230°C indicates the iron is completely molten. The *bath* continues to rise in temperature until it reaches the temperature of the furnace. This is called *equilibrium* (about 1,425°C or 2,600°F). To this point, the vertical line on the phase diagram simply tells you that the iron, when it was charged into the furnace, was a solid at room temperature. When it reached the temperature of the solidus line, it began to melt. Between this point and where it reached the temperature of the liquidus line, it was a mixture of varying amounts of liquid iron and solid iron. Above the liquidus line, the iron was always all molten. It just got hotter until it reached equilibrium with the furnace temperature. We can almost say that this vertical line represents the work of the puddler's helper. He charged the furnace with raw material, and he shoveled the coal and stoked the fire to heat the iron until it all became molten. The work and skill of the puddler is represented by the horizontal line. As he used the excess oxygen from the flame of the fire and also that contained in the mill scale or some ore, the mixture began to contain less and less carbon (moving left along the line) even as the temperature remained the same. At the point where the horizontal line reached the liquidus line, solid bits of pure iron started to precipitate from the molten bath. This continued to happen as the percentage of carbon became less and less and the puddler continued to stir and gather the solids into an increasing number of larger and larger balls—until the point where the mass reached the *solid* line on the graph. At this time the furnace temperature was not high enough to maintain any part of it in the liquid (molten) state because it was too pure to exist in that form at that temperature. Note: the graph shows that pure iron with no carbon requires a temperature of about 1,535°C to be maintained in a molten state, or about 125°C higher than a puddling furnace is capable of generating. The "magic" of the puddling furnace was simply to reduce the amount of carbon in a molten bath of iron. As the iron became more pure, the furnace was not hot enough to maintain it in the liquid state, and it would then become a solid of its own volition. Balancing that was the puddler's skill in preventing reoxidation of the iron. (For photos of various aspects of puddling operations, see figures 1.3 through 1.7.)

Figure 1.3. Shows the puddler who is rabbling the "heat" through the working door of the furnace.

Photo, Colonial Steel Co. Pittsburgh. Collection of the Author.

Figure 1.4. Slag flowing from the furnace during the "boil."
Photo, Colonial Steel Co. Pittsburgh. Collection of the Author.

Figure 1.5. Removing the "puddle ball" from the furnace.
Photo, Colonial Steel Co. Pittsburgh. Collection of the Author.

Figure 1.6. After it was removed from the furnace, the "puddle ball" was inserted into the jaws of a "squeezer" where much of the slag was removed and the metal was formed into a shape called a "bloom" that could fit into a rolling mill.
Keystone View Company, Collection of the Author.

Making wrought iron was a small-batch process. As previously stated, only a few thousand pounds were produced in one furnace in a day, and that in four or five smaller lots or heats. The material was thought to exhibit a number of excellent qualities, especially corrosion resistance. It became even more important because it was the source of raw material for the crucible steelmaking process. Because of this niche, it continued to be manufactured well into the era of mass-produced steel.

In relatively recent times (1930s through the 1960s), the A. M. Byers Company of Pittsburgh produced a pseudo form of wrought iron, made in relatively large quantities by the Aston process. Iron was decarburized (reduced in carbon content) in a Bessemer converter, twenty to twenty-five

Figure 1.7. An operator inserts a bloom between the rolls of a rolling mill while another is waiting on the conveyor from the squeezer to the right. Note that a "red hot" strip of "muck bar" is exiting the rolls to the lower left of the operator.
Photo, Colonial Steel Co. Pittsburgh. Collection of the Author.

tons at a time (to be described in a later chapter). The refined iron was poured into (through) a ladle of clean, molten siliceous (iron silicate) slag. As the stream of metal flowed through the slag, it formed into little globules, because the slag was maintained below the "freezing" temperature of the refined metal. Encased in slag, the metal collected on the bottom of the slag ladle was in an irregular blob, similar to the previously described puddle ball. When the pouring of the metal was complete, the slag was drained off for reuse, and a pile of slag-laden metal, called a "sponge ball," was dumped into a large nine-hundred-ton squeezer to remove the excess slag. The bloom was then sent to the rolling mill and converted into useful products, principally pipe. Production rates for this "modern" process were about six thousand to eight thousand pounds per ball every five minutes,[29] or about forty to fifty tons per hour.

Chapter Two

Crucible Steel

Crucible or *cast steel* was invented by Benjamin Huntsman around 1740 in the Sheffield (Doncaster) region of Britain. Huntsman was a clockmaker, and the customary story is that he created crucible steel in pursuit of a suitable quality material from which to make springs for his clock-making enterprise. However, in all likelihood this story is not true.[1] It should also be said that the story of steel does not begin here; steel existed before Huntsman, in both crude and elegant forms, which is why we must always use qualifiers such as *cast* and/or *crucible* with his name. It is also the reason that it is more proper to say that he "rediscovered" quality steelmaking instead of inventing it.

The story of steel essentially begins with ancient *wootz* and *Damascus* steels. Wootz steel was a product of ancient India and was probably the source for the elegant Damascus steel swords as well. They were noted for their sharpness, hardness, and elegant quality, as well as for the art of manufacture that was lost to European ideology in the intervening centuries. John Percy's description of wootz follows:

> Little lumps [of iron] produced in an ordinary Hindoo iron furnace, is put into a conical unbaked clay crucible, of about the capacity of a pint, with the addition of . . . [Dried wood and two leaves.] The crucible thus charged is closed at the mouth with a cap of unbaked clay well luted on, and then well dried near a fire. The furnace used is a little circular pit in the ground somewhat dilated at the top. An earthen pipe, connected with two bellows . . . which are worked alternately, enters the fireplace at the bottom. . . . The fuel is charcoal, with which the fireplace is filled . . . is covered over. The bellows are then plied during 4 hours, when the operation is completed. . . . On breaking open the crucibles [after cooling] the steel is found melted into the well-known conical shape of wootz. . . . The steel in each crucible is reckoned by workmen to weigh . . . somewhat more than 2½ lbs.[2]

Damascus steel was occasionally called *Spanish* steel, reflecting the Moorish influences on Spain. Roman swords and weapons were made of iron, often interspersed with strings of steel created by the forging process. Roman conquests are sometimes attributed in part to the ability of their weapons to cut through and destroy the softer bronze and copper swords, and others, of those they conquered. One can only imagine what they might have done had they been acquainted with the secrets of steel.

Two of the more crude forms of steel that existed in Huntsman's time were known as *blister* or *cementation* steel and *shear* steel, although there were a few other types. Blister steel was made in a cementation furnace by stacking piles of wrought iron bars (usually called *muck bar* in the United States) into a refractory box or boxes located within a furnace. The layers of bars would be completely surrounded by one-half inch to one-and-one-half inches of finely ground wood charcoal or some similar material (the *cement*). A bar was not permitted to be in contact with another. After the box was filled, it was covered with some previously used cement over which was placed a layer of refractory powder thick enough to prevent the furnace gasses from penetrating to the bars. The furnace would then be closed and sealed, and heating was started. It would typically be kept burning for seven to eleven days.[3] The fire of the furnace would flow around the outside of the metal laden boxes, transferring its heat to them. This would cause some of the carbon from the incandescent charcoal cement in the box to diffuse through the iron and into the bar. The amount of carbon content in the bar was in direct proportion to both the temperature and duration of time that the wrought iron was in the furnace.[4]

Cementation or blister steel takes its name from bubbles on the surface of the bar that resembled blisters on a person's skin. Recall that wrought iron had very little carbon content, so to give it more tenacity it was necessary to partially recarburize it. Today, some steels have much lower carbon content—so low that it is measured in parts per million versus parts per thousand for wrought iron. Wrought iron also contained entrained slag. This would be unacceptable in very low carbon or any modern steel. Refer to the phase diagram in figure 1.2. Had the puddler the knowledge of when the carbon content was correct, and had he been able to stop the process in the range classified as steel (top ordinate on the phase diagram), then no further processing would have been necessary. Parry and other sources claim that this was routinely done at the Ebbw Vale Works in Wales.[5] If costs and skills were reasonable, why didn't it become a common practice elsewhere? How could they have avoided the entrainment of slag? Cast iron was hard but very brittle due to its high carbon content, while wrought iron was soft and very ductile because the carbon content was very low. Making steel involved that happy medium lying somewhere in between the two, so that it met the end-use requirement. There is not, nor ever was just one, single chemistry of an alloy of iron that defines steel.

The quality of the product produced in a cementation furnace was quite varied. To make the bar more homogeneous, a material called *shear steel* was developed. To produce shear steel, short bars of blister steel were stacked together and clamped on one end, while the opposite end was inserted into the forge fire. On reaching welding temperature, the stack was taken to the hammer and welded together. Then the opposite end of the stack was heated and hammered. The welded bars were frequently stacked together a second or even a third time, and the procedure was repeated.[6] This more uniform steel was probably the material that Huntsman would have selected to make clock springs. Some version of the cementation process continued well into the era of modern steelmaking as a method of "case hardening" steel.[7]

Experience from making shear steel surely taught Huntsman that a goal in manufacturing quality steel was making a homogenous mixture.[8] If you had not been exposed to the three steps in making shear steel—first produce wrought iron, then blister steel, and finally make shear steel—the next statement might seem absurd. The raw material that Huntsman used for making steel was steel. He used blister steel, placed in a covered crucible and heated in a coke-fired furnace, to make what has become known as cast steel. He made the best quality steel, but he kept his process as secret as possible. Consequently, little is known of his actual operations. What we do know is that the crucibles he used were made with local clays that were heated in a coke-fired furnace. Using coke was advantageous, because it burns hotter than coal or charcoal (1,530 to 1,600°C for coke versus 1,425°C for charcoal, although the hotter burning range of coke probably wasn't available in Huntsman's era).[9] The use of blister steel as a raw material was advantageous. Again, recall the phase diagram. Blister steel contains more carbon than wrought iron and therefore has a *lower* melting temperature. Maintaining the furnace temperature became less critical than that associated with making wrought iron because coke burns hotter than coal. Furthermore, the crucibles were imbedded in the burning coke, whereas with coal in an open furnace, the metal was exposed to the oxidizing flame and all of the extra air needed to burn the coal. Even so, he was operating near the limit of his ability to keep the metal molten. His ingots were not large. Early in his work they weighed no more than a few ounces, and in 1761, not many years before his death, only thirteen pounds.[10] Yet they commanded a high price.

Even though Huntsman tried to maintain secrecy, others began producing cast steel. Sheffield became the quality steel-producing center of the world. A number of decades after Huntsman's death, someone began producing cast steel directly from wrought iron, thus saving the step and cost of first producing blister steel. This change was in fact heralded some two decades before Huntsman developed quality steel. A Frenchman named Reaumur experimented with carburizing wrought iron with pig iron.[11] In 1800 David Mushet was granted a patent for the "manufacture of cast steel, by the fusion of bar or wrought iron with carbonaceous matter [an example, the black oxide of manganese] in crucibles."[12] This required the addition of a form of carbon directly into the pot, negating the need for blister steel. As you will read, this advance will have an immense impact in the future history of steelmaking during the era of *mass-produced steel*.

An important aspect of the Sheffield domination of this industry involved the development and production of quality crucibles. Crucibles played an important role in the process of making steel, much more than the obvious function of a container. They had to withstand being held at 1,500 to 1,600°C (2,700 to 2,900°F) for four hours and longer, enduring somewhat rough handling, and they had to resist bursting while holding sixty to one hundred pounds of molten metal. Crucibles participated to some extent in improving the chemistry of the steel but, if not controlled properly, they could make it worse. They needed to resist the eroding effects of the slag and be capable of repetitive use.[13] These demands made the cost of crucibles a rather large portion of a crucible steelmakers' expense and had to be controlled. This cost and the limited access to special clays effectively prevented competition from outside of the Sheffield sphere of influence.[14] Though these crucibles did not fare well during transport, there were British laws that made their export prohibitive.

That is not to say that steelmaking wasn't being attempted elsewhere. There were many failed efforts in New York, New Jersey, eastern Pennsylvania, and other regions, as well as outside of America. In Pittsburgh, Simeon Broadmeadow successfully manufactured blister steel between 1828 and 1830, but it appears that cast steel made from it was unacceptable.[15] The first successful crucible steel made in the United States was produced in Cincinnati in 1832, but the company failed for economic reasons by 1837. The Shoenbergers had been making cementation steel in Pittsburgh since 1833 and attempted to manufacture crucible steel in 1840, but failed because "the crucibles employed were made of American clay, and, as may be supposed, were ill suited for the purpose required."[16] Joseph Dixon failed at his attempt to make steel (1827) but later developed a crucible using Ceylon graphite that overcame such deficiencies. These compared favorably to Sheffield pots, so he started a successful company in New Jersey to sell them.[17] Crucibles containing graphite were usually called *plumbago* crucibles and became a standard in the United States. Between 1845 and 1852 a number of other Pittsburgh firms were successful in making cast steel, though not always the best quality. "[The] first slab of cast steel rolled in this country had been shipped from Pittsburgh to the Deere Plow Works at Moline in 1846."[18] In 1853 the Singer and Nimick Company made saw and machinery grades regularly, but not tool steel. Finally, in 1860 Hussey, Wells & Co. is considered the first company to produce the best quality routinely.[19] "By 1860 the amount made in America had nearly doubled, and almost one-third of the 12,000 tons we then produced were cast or rolled. More than half of the country's product was made in Pittsburgh."[20] Pittsburgh had thus arrived for the first time as the steelmaking center of United States, notwithstanding the fact that until the previous year no iron was smelted in the area. All of it had been imported from other regions. Overcoming the frequently undeserved mystique of Sheffield superiority contributed to this success. The failure of the American government to provide a protective tariff to help sustain this important evolving industry was another obstacle. Blockades during the American Civil War probably contributed more than any other factor by forcing the use of Pittsburgh and other steel-producing area's products to expel any sense of American inadequacy.[21]

The process of making crucible steel is often described as a simple two- or three-step process, defined as *melting, killing, and teeming*. Melting is just the span of time during which the cold charge of metal is heated until it becomes molten. Killing is the period of "holding it molten in the crucible, which still remains in the melting hole [of the furnace], some change occurs which removes the tendency to form blowholes."[22] Finally, teeming is merely the act of pouring the molten metal from the crucible into an ingot mould.

The process evolved somewhat from Huntsman's time. Hotter-burning coke made in beehive ovens permitted the use of wrought iron instead of blister steel as a feedstock. Adding carbon directly to the pot in the form of charcoal, pig iron, or some other "carbonaceous" matter (usually some form of iron-carbon alloy), proposed by Mushet, became the norm when using wrought iron instead of cemented steel. Scrap steel was sometimes used, but almost without exception it was cited as being responsible for an inferior product.[23] American and British practice began to diverge somewhat with the introduction of the plumbago crucibles. The graphite that was used to make the pots is a form of carbon, and so the pot itself contributed some part of the carbon needed to convert the wrought iron into steel. In Sheffield they continued to rely primarily on clay pots, and for a long time they continued using blister steel. Americans preferred using newly developed gas-fired furnaces, whereas the Brits used coke. In America the crucibles were usually filled while they were cold, while in England when they were hot. Pittsburgh men let the crucibles cool and inspected them prior to each reuse, sometimes for as many as ten or more cycles. Sheffield workers typically used a clay pot three times, never letting it cool, and reduced the charge weight (and therefore volume) of each successive charge. The successively smaller charges were due to the cutting effect that the slag had on the pot. The slag line needed to be lowered. Also, clay pots would frequently fracture if left to cool, being one of the big advantages of using plumbago. Still, the basic process remained substantially the same.

"From a chemical standpoint, at least, the crucible is the simplest of all the processes for making steel."[24] While crucible steelmaking has been portrayed as being simple by description and in concept, it is anything but in actual practice. It has been a method that tested the limits of strength and the endurance of men. Because this story is about men, the work they did should be understood. Because this type of steel was made in England and in America (not forgetting Europe and elsewhere) and because it was made in coke-fired as well as gas-fired furnaces (anthracite, too), there were slight variations in the titles of the assorted jobs that men were called on to do. In America the leader/boss was called the *melter* and the next in line of importance was the *puller-out.* Lesser workers were the *setter-in* and the *pot-packer* on gas-fired furnaces. On furnaces that used anthracite coal or where the fuel was producer gas, there was a *coal-wheeler.* In Sheffield there was the *teemer* or *melter*, *puller-out*, *odd-man*, and *cellar-lad* on the coke furnaces.

A crucible furnace looked somewhat like a platform or a stage, slightly raised above ground level. It appeared as though it was an integral part of the building in which it was housed, and many were paved in cast iron plates. Inserted in groups into the floor of the stage were the melting holes that held the crucibles, frequently two or four or six in each hole. This was where the heating occurred. There were removable covers, edged by a brilliant light emerging from the inferno below (see figure 2.3). In a gas furnace, typical in the United States, the full pot would be charged cold by the setter-in onto six to eight inches of coke dust lining the floor for protection. The crucible was filled with cut bar by the pot-packer (see figures 2.1 and 2.2), who first added the charcoal or other carbon form desired. This was needed to make the conversion from wrought iron into steel. Along with the carbon, some other materials that were thought to promote fluxing, a fluid slag or something similar, was added. It got to the point that they were using "so many different substances or 'physics' and mixtures that the stock house came to be known as the medicine house."[25] In coke or anthracite furnaces, the odd-man inserted the pots back into the coke/coal while empty but still hot. Then coke or coal was filled in around them. Care had to be taken so that at no time during the steelmaking process was a piece of coal or coke dropped into the pot. Even a small chunk could add enough carbon to ruin the entire lot. The chilled metal from this kind of abortive heat was said to be "*hot short* and *stares*, i.e. it has a splendent fracture."[26] These pots were filled by the odd-man while the puller-out held a funnel to guide the metal charge into the vessel by a process sometimes referred to as *steeling* (see figure 2.8).

The loaded crucibles were heated during this *melting stage* (see figure 2.4). It typically lasted for three to four hours depending on a number of factors. Low carbon heats required more time than high carbon heats, first heats of the week required more time, and so forth. This was a somewhat quiet time for the men. Gas furnaces would have the heating direction reversed, typically every half-hour during this period. Coke holes had to be shaken down and the surrounding coke filled to near the top of the crucible. This wasn't necessary on an anthracite furnace. Men frequently used this time to make up the water lost through sweating by imbibing some of the product of a local tavern. In England, the search for a suitable stock was in the charge of the cellar-lad. When the melter (teemer in the United Kingdom) thought that the metal was melted, he would pull the lid off of each crucible, and he could judge the temperature by eye, but sometimes, if he were unsure that the entire pot was melted, he would stick a thin, metal rod into the pot to feel around. Looking at molten metal requires some skill; one must be able to stare into the intense and brilliant light and adjust his eyes to be able to inspect the shimmering surface of the steel, not a talent that everyone can command. If it was melted, the *killing phase* began. This typically lasted for a half-hour to an hour, more toward the shorter time in the United States. It was during killing that expanding bubbles of gas called *cat's eyes* would be seen and then burst. Sometimes there would be a small spurt of sparks. Observing these phenomena told the melter that the heat was not yet killed. This was one of the instances in which the composition of the crucible played a part in affecting the chemistry of the steel. The first was by its contribution to the carbon in the heat. This had been anticipated and accounted for when the charcoal or ferromanganese was added to the pot, if plumbago pots were used. Second, silica from the clay in the pot formed silicon that was responsible for stopping the evolution of gasses in the molten steel or what was defined as killing. If not enough silicon entered the cast steel, the finished ingot would be full of blowholes. If too much silicon entered by holding it in the pot for too long, it became hard and brittle and was ruined.[27] So the steel had to be carefully observed during this period. In the latter era of crucible steelmaking, if there was still a question as to whether the pot was killed, a *pill* of aluminum was added to ensure that, in American parlance, the steel was *dead*. If a small sample of molten steel was poured into an open mould before it was killed, one could actually view the volume of metal increase or "grow" due to the outward pressure of the expanding gasses within. This makes it easier to understand the jargon used by the men.

When the melter declared that the steel was killed, the truly heroic segment of crucible steelmaking began. The gas was shut off or the coal or coke was run down. The puller-out, clad in wooden shoes, his legs and arms wrapped in layers of soaking wet burlap or similar protection, would then squat over the melting hole with his legs spanning the opening like they were the beams of a bridge. His entire body was exposed to the inferno as if he were a sacrifice to the fire. He then had to stare through the blinding light of the hole to find the pot (see figure 2.5). To cite the often-quoted Harry Brearley, simply because his is the finest description to be found:

> At steel melting heat the pot is not "as hard as a brick" . . . The pot is soft and in a degree yielding: it could be hit with a hammer and deformed without cracking. . . . The puller-out can feel the "give" of the pot when his tongs grip it. . . . [Furthermore,] It goes without saying that a man that can lift a pot containing sixty pounds of molten steel with a pair of heavy tongs from a furnace below ground level at a dazzling white heat is no weakling. I say "lift" but the pot is not lifted; to call the men "lifters" instead of "pullers-out" would be insulting. The actual pulling out is like Macbeth's job: "When 'tis done, then 'twere well it were done quickly . . ." [Then,] The feeling of "give" gives him the confidence to straighten his back and with an unbroken pull and swing to set the pot on the floor-plates.[28]

Occasionally the puller-out was treated to the surprise of a broken pot, which was accompanied by excruciating pain conveyed to him by the sixty to one hundred pounds of nearly three-thousand-degree molten metal playing out beneath his fully exposed body. The entire operation of pulling took perhaps fifteen seconds for one pot. For an entire day the act of pulling pots was repeated some twenty to forty times, and was therefore only measured as a matter of a few minutes. This was likely an eternity for a lesser man. Still, the job was not yet done, as the puller-out "then swings it across to the teeming-hole, close to the ingot mould to be filled"[29] (see figure 2.6). "The difficulty of the work is attested by the fact that one pull consists of the crucible, 40 pounds; charge, 100 pounds; cap, 5 pounds; tongs 20 pounds; a total of 160."[30] Fuel usage rates were about one ton of coal per ton of steel on producer gas-fired furnaces and three to four tons of coke per ton of steel on coke furnaces.[31]

Continuing with Brearley's observations:

> The teemer [melter] is the autocrat of the furnace gang and the best paid man of the lot; . . . As he has been a puller-out in his time, the teemer is generally big and strong, but as he is no longer young he may be rather fat, and has generally a ruddy face. The colour of his face may be due to a combination of good health, the scorching heat of fires, and the stimulation of alcoholic drinks . . .

> Except when weighing up the charges for the next heat the teemer has only a supervisory job until the steel is melted and ready for casting; then he is the Shah of Persia, . . . Like the puller-out, the lower parts of his legs are protected by sacking laid on to a foundation of tarpaulin or some other waterproof stuff. These trappings are called "rags," . . .
>
> As soon as the puller-out has swung the pot of molten steel up to the ingot mould and removed the lid, one of the odd men, or the cellar-lad if he is lucky, inserts a slim rod with a blob of metal at the end of it about the size of an orange, and mops off the floating layer of slag; . . . During this interlude the teemer has been holding the pot in his bowed tongs, . . . Having mopped off the slag, the molten steel is visible. What there is to see distinguishing one pot of steel from another the novice cannot tell. But the teemer can. If ever so few sparks should rise from the surface the old melter would grunt and mutter some doubt about the steel being killed; . . .
>
> The teemer grumbles if he considers the molten steel to be on the cold side, as that limits his choice of moment and manner of teeming; to hurry molten steel into the ingot mould for fear that it will not get there at all is considered a disgrace . . . Watch him teem; and if the physical grace of a bulky man and the play of colour around the pot do not enchant you, try to realize what it is that he is trying to do . . . From the bowed end of the tongs which grip the pot, long shanks pass over the knee of the crouching teemer and extend as far as his left arm can comfortably reach. With the leverage of the long shanks operated by one hand and a steadying lift exerted by the other, the pot is balanced on the teemer's knee at the moment his good judgment decides is the right moment for casting. The metal flows over the lip of the pot, . . . passes down to the mould, and from the first to the last, at whatever speed it may be delivered, does not make contact with the side of the mould. No lady, handling a delicate china cup, ever sipped tea with greater niceness than the knowing melter delivers the glistening stream of molten steel into the soot-lined mould.
>
> . . . But the mould does not stand vertically: it slopes toward the falling stream of metal at an angle . . . Unless these trifles are considered, no amount of skill could deliver the steel into the mould without "catching" its sides; and a "catched" ingot is, on all faults, the most obvious sign of negligence or incompetence.[32] (See figure 2.7.)

In America some manufacturers started using considerably larger crucibles than those in England, which held a charge of up to 160 pounds by Barraclough's account and 200 pounds according to Howe. This would equate to men lifting approximately 225 to 265 pounds per pull. Taking into consideration all of the other adverse conditions, these pots certainly could not be either pulled or teemed by hand, and to the consternation of the traditionalists they were pulled from the hole by crane, wheeled to the mould, and teemed by a method that some called *tipping* (see figure 2.9) or by use of a special set of tongs that required two people to lift the pot. In the United States the routine had been developed to teem into a ladle (see figure 2.10). This had two benefits: larger ingots could be made from a number of pots, and variations that were found in the steel's chemistry from pot to pot were diminished due to the blending that occurred in the ladle. Since multiple crucible casts had been done for special reasons for decades, this was nothing new. Sheffield men called tipping "sloppy." Tipping was simply using the metal tongs to lean the pot over on one edge and teem while not lifting it off of the floor. It did not require the strength or finesse of the customary method. It is easy to understand how an "artisan" might feel.

Ingot moulds were made of iron and were manufactured in two pieces that formed a unit that was held together with rings and wedges. The inside surfaces were cleaned and smoked, so that it had a thin layer of soot to prevent the molten metal from sticking. Once the metal had cooled and solidified, the moulds were removed and the ingots were inspected (see figure 2.11). The solidification process was started when the molten metal was chilled by contact with the surface of the mould. This process continued from the outside to the inside. Because the metal was shrinking as it cooled, the inner molten metal acted as a reservoir for the rest of the ingot, and a "v" or conical-shaped void called the *pipe* would form in the top portion of the ingot. After it was totally cooled and the mould was removed, the ingot would be *topped* by knocking off the offending section of pipe. This was continued until the pipe was removed (20 percent or more was not uncommon). This had to be done because the irregular-shaped, oxidized surface of the pipe would not weld together when heated and hammered or rolled. If not addressed properly, it would form a weakness in the finished product. Numerous methods were developed by steelmakers to address the problem of piping in ingots. This was a large source of production loss and a problem that lasted into the modern era of steelmaking in the final quarter of the twentieth century, when continuous casting came into common use. Small numbers of ingots are still cast to this day.

In many of the American shops the melter was paid for the work, and he would in turn pay the entire crew. In Pittsburgh, this gave power to these skilled crafts and continued to favor them in their negotiations with those on the capital side of the ledger. In one shop butchers, bakers, and other unskilled workers were hired for the skilled positions, based on the fact that the "crucible is the simplest of all the processes for making steel." Apparently, the amount of ruined steel was high. The crucible process compounded the need for and the importance of the puddlers, who supplied the raw material for the cast steel makers. This was a heady era for labor in steel, and it continued in command.

Electric furnaces that were developed in France in the 1890s and shortly thereafter (about 1906) introduced into the United States gradually replaced crucible or quality steelmaking. Electric furnace steelmaking was somewhat like crucible steelmaking in that the principal function was melting, although a small fraction of carbon was added to the heat through the consumption of the carbon electrodes. The benefits were the reduction of labor and exposure to the extreme elements of the crucible steelmaking process. Production rose to a number of tons per charge instead of the previous one-hundred-pound lots. Even so, crucible steelmaking continued in the United States until the World War II era, proving that old ideas die hard.

Figure 2.1. Shearing wrought iron muck bar for feed stock for the crucible steelmaking process.

Photo Colonial Steel Co. Pittsburgh, circa 1912. Collection of the author.

Figure 2.2. Pot-packers filling crucibles for the steelmaking process.
Photo Colonial Steel Co. Pittsburgh, circa 1912. Collection of the author.

Figure 2.3. View appears to be a four-hole gas-fired crucible steelmaking furnace. The curved bar suspended from the chain is a lifting device for removing the covers from the furnace holes. Coke fines and spare crucibles line the walls.
Crucible Steel Co., LaBelle Steel Works, Pittsburgh. With permission ASM International, Metal Progress, Crucible Steel–Made in America, May 1940, Van Fisher photo.

Figure 2.4. Six loaded crucibles with lids in a melting hole.

Crucible Steel Co., LaBelle Steel Works, Pittsburgh. With permission ASM International, Metal Progress, Crucible Steel–Made in America, May 1940, Van Fisher photo.

Figure 2.5. The puller-out, poised over the melting hole, preparing to pull a crucible from the melting-hole in the inferno below. It is so bright that it is difficult to see inside the furnace.

Crucible Steel Co., LaBelle Steel Works, Pittsburgh. With permission ASM International, Metal Progress, Crucible Steel–Made in America, May 1940, Van Fisher photo.

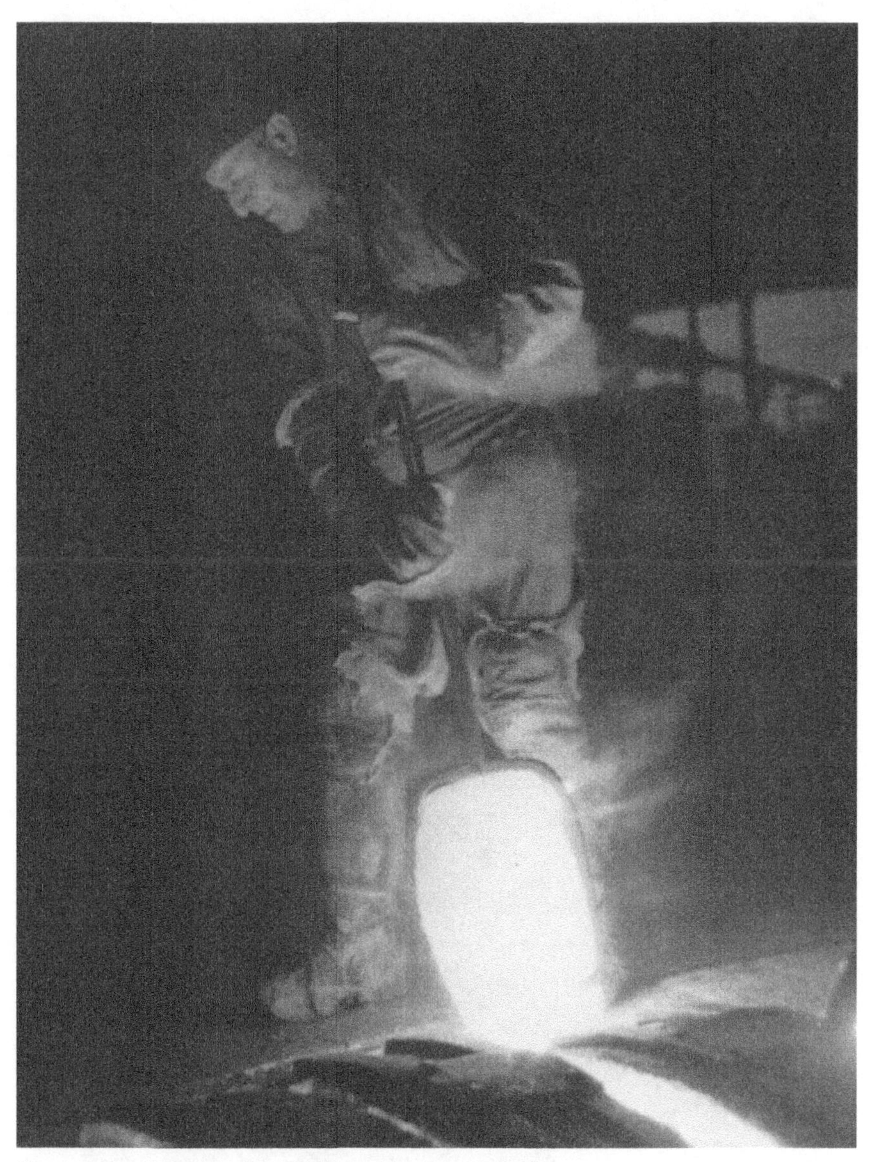

Figure 2.6. Swinging the crucible to the teeming-hole.
Smithsonian Institution collection.

Figure 2.7. The melter (teemer in the UK) teeming a crucible of molten steel into an ingot mould. Note the bend in the tongs, suggesting the heavy weight.
Photo circa 1925, Crucible Steel Co., Carnegie library of Pittsburgh collection.

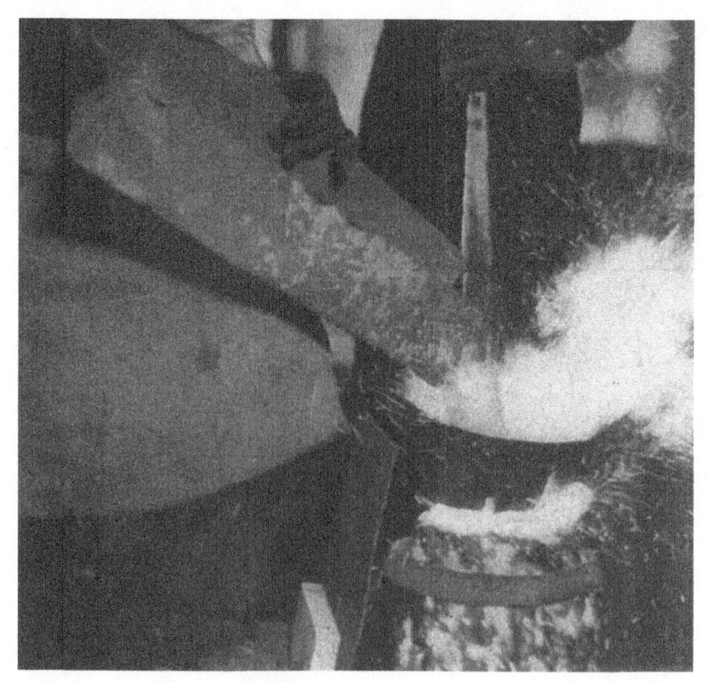

Figure 2.8. Filling a hot crucible with a new charge of sheared muck bar by a process sometimes referred to as steeling.
Crucible Steel Co., LaBelle Steel Works, Pittsburgh. With permission ASM International, Metal Progress, Crucible Steel–Made in America, May 1940, Van Fisher photo.

Figure 2.9. Teeming an ingot by "tipping," which was considered to be sloppy by artisans.

Photo, Colonial Steel Co., Pittsburgh. Collection of the author.

Figure 2.10. Multiple crucible heats were sometimes poured into a ladle in order to form a larger ingot than was capable from teeming a single crucible.
Photo, Colonial Steel Co., Pittsburgh. Collection of the author.

Figure 2.11. Stripping the split-mould from a large crucible steel ingot.
Photo, Colonial Steel Co., Pittsburgh. Collection of the author.

Part II

THE BEGINNINGS OF SMELTING— A TRINITY OF IRON

Chapter Three

Fuels and Transportation

COAL

Undoubtedly, one of the essential elements in Pittsburgh's rise to become one of the world's greatest manufacturing centers, similar to that of Sheffield, was the availability of fuel. The fact that the city and the region contained vast resources of coal that were exploited early in its history is well documented. According to Howard Eavenson, a distinguished author on the history of the industry, coal was mined in Pittsburgh during colonial times. He relates that in a note to Colonel Bouquet, the immediate successor to General Forbes as the commander at Fort Pitt, from Colonel Mercer, Bouquet's replacement at Fort Pitt, Mercer indicated that the soldiers of the garrison were mining coal in the vicinity of the fort as early as 1759, and that they might have even been converting it into coal tar to replace pine tar that was in critically short supply. Tar was used as a lubricant for wagon axles.[1] The geologic history of the region had been exposed through the erosion of the Allegheny plateau by a series of intersecting rivers and streams contained within their steep valleys. This history showed a unique regional advantage, a fact that is not immediately apparent to the casual observer. In 1788, Dr. Johann David Schoepf referred to Coal Hill, or what is present-day Mt. Washington, and commented: "The coal bed mentioned midway of the hill or mountain, is so much more noteworthy because elsewhere coals are to be dug for at depth, and is proof of what great changes have taken place in the surface of this region . . . The vein of these coals is 10-12-18 ft. wide and extends throughout the length of the mountain."[2] Victor Collot attested to the quality of the product in 1796, and proclaimed: "A rich vein of coal is found on the summit of one of the mountains which bounds the Ohio on the left. The quality of this coal is equal to the best kind in England; the mine is open, and the coal is so cheap, and forming such an excellent fuel, that although the inhabitants live in the midst of forests, they prefer it to their best wood. It costs less then four-pence sterling a bushel."[3] The coal was exposed for all to see and exploit. It was found in thick seams where a man could stand tall or nearly so while mining. These types of mines were called *drifts*, where the coal was attacked sideways into the hill, as opposed to other techniques referred to as *slope* and *shaft mining*, where an access port had to be dug down to the coal seam before mining could begin. Recovery of the product was easier as well, as there was no need to hoist coal or waste up a vertical shaft, or pull it up a long, sloping incline to the entrance portal. Pooling of water and/or drainage of same was not a typical problem in drift mines, where this was often critical in deep mines.

Pittsburgh used this to its advantage almost from the beginning. As with many advantages, there were shortcomings. From an article in the *Navigator* of 1811:

> On entering the town the stranger is rather offended with its dark and heavy appearance. This arises from the smoke of the coal, which is used as the common fuel; and of which about 250,000 bushels are consumed annually. It costs about six cents a bushel delivered at your door, and is said to be equal to any in the world. Our rough hills are filled with it, and our rooms in winter feel the effects of its warmth, and cheerfulness. Wood as an article of fuel costs 2 dollars a cord delivered—The abundance and cheapness of coal will be peculiarly advantageous to Pittsburg in her progress in arts and manufactures. Coal Hill, on the south side of the Monongahela, abounds in coal; and a pit in it is said to have taken fire about the year 1765, and continued burning for 8 years; and another pit on Pike run, which burned for 10 years. This is a high and steep hill, and its top affords a handsome prospect of the town and rivers below it.
>
> From the immense quantity of coal burnt, there arises a cloud of smoke which hangs over the town in a body, and may be seen at two or three miles distance; when in the town, this cloud of smoke is not discovered, and the place soon becomes familiar to the eye, while the ear is occupied with the mixed sounds of the implements of industry from 5 o'clock in the morning till 9 at night.[4]

Additionally, T. M. Harris spoke of problems with drinking water from some of the springs in Pittsburgh during 1803: "But the spring water, issuing through the fissures in the hills, which are only masses of coal, is so impregnated with bituminous and sulfurous particles as to be frequently nauseous to the taste and prejudicial to the health."[5]

Deeply rooted in history like the London smog, this image of a polluted Pittsburgh has yet to abate. Its air and water have been as good as or cleaner than most major American cities for decades.

The manufacture of wrought iron and cast steel were intensely fuel dependent. Puddling required at least a ton of coal for each ton of wrought iron made. Likewise, crucible steel needed one ton of coal for each ton of product when using a producer gas-fired furnace. Producer gas was made from coal. The coal-equivalent requirement for a coke-fired crucible steel plant was even higher. About three or even four tons of coke were used to make one ton of steel, and it took in the range of 1.5 to 2 tons of coal to make one ton of coke at this point, depending on the method of coking. This equates to about four-and-one-half to eight tons of coal per ton of steel. These figures did not include the amount of charcoal or anthracite coal needed to make the pig iron, nor the charcoal or other fuel necessary for creating blister steel, if such was the feedstock used for the crucible. Also, this does not account for the coal used in the forge fires or the boilers that powered the hammers or rolling mills. It took about one acre of hardwood, converted into charcoal, for every ton of pig or cast iron made. Because all pig iron was imported into Pittsburgh until ironmaking was inaugurated in late 1859, this was not a problem for the area. Also, when smelting did begin here it was a coke-based process. At that time coke-fired blast furnaces were uncommon in the United States, but that is getting a little ahead of the story. Importing iron was easier and cheaper than exporting the coal or coke needed to make it, even when it was possible, as charcoal or anthracite fuel was the norm and coke-fired blast furnaces in this country were rare or very nearly nonexistent. Consider that it was largely an uphill haul from Pittsburgh to the furnaces in the east or north and required a larger volume too—more than a ton of coal or coke was needed for every ton of iron made where coke furnaces were in operation. Conversely, since iron is much more dense than coal or coke, it was therefore more compact. Additionally, the trip to the city from the mountains was more or less downhill as well.[6] Then, too, there would have been the necessity of securing and delivering supplies of iron ore. At this time, these facts certainly favored imports.

The Pennsylvania Historical and Museum Commission described the area's resource of coal, named the Pittsburgh Seam, as being "this State's [Pennsylvania] bituminous coal industry . . . born about 1760 . . . eventually judged the most valuable individual mineral deposit in the U.S."[7] Also, as depicted in 1868, "to contain fifty-four billion tons. Locally this was valued at more than the estimated output of the California gold fields for one thousand years."[8] While the latter citation is an exaggeration, the former is not. The Pittsburgh region had other formidable coal seams that were not included in this calculation; the Freeport Seam is one example. The Pittsburgh seam extended for some six thousand square miles,[9] typically varying in thickness locally from six to eleven feet and to a maximum of fourteen feet in the Piedmont.[10] That portion of the Pittsburgh Region, an area about fifty miles square (2,500 square miles) confined within the borders of the state, was estimated

Table 3.1. Bituminous coal production in Pennsylvania for selected years. Note that output was in essence *doubling* every ten years until 1920. Also, around 1810 other coal seams began contributing to the total. Although these other coals were becoming more important, the Pittsburgh Seam's contribution was the controlling factor.

Year of Production	*Pittsburgh Seam (Tons)*	*All Bituminous in PA (Tons)*
1790	30,000	30,000
1800	87,000	87,000
1810	120,000	120,700
1820	200,000	225,600
1830	300,000	397,700
1840	410,093	699,994
1850	1,317,000	2,147,500
1860	2,400,000	4,170,400
1870	4,630,000	9,223,856
1880	11,900,000	16,564,440
1890	*	40,784,003
1900	*	79,318,362
1910	*	148,770,858
1920	*	166,929,002

* no data

Source: Howard N. Eavenson, *The First Century and a Quarter of American Coal Industry* (Pittsburgh: Privately Printed, 1942), 450–51; Clyde H. Maize and George S. Struble, *History of Pennsylvania Bituminous Coal*, Pennsylvania Department of Mines, 1955, 58–59.

by the 1886 Geological Survey to contain ten billion tons of remaining coal[11] (see table 3.1). Yet, stunning as these figures may be, there was a very special section of this coal bed, about three miles wide and sixty miles in length, known as the Connellsville District[12] after one of the principal towns in the area, the product of which would stagger the world. Later discoveries would extend this special region to about eighty miles.

COKE

Coals are typically categorized by a number of different material qualities. One important characteristic is called the *volatile matter* content. Although coal looks like a rock, part of it will soften and liquefy when heated, like a cook does to sugar when heating it in a saucepan (caramelizing). The amount of this volatile matter contained within the coal, named as such because it vaporizes if heated to a high enough temperature, is a means of ranking coals. If coal was burned—that is, the volatiles and carbon were heated in the presence of air and a source of ignition—the only thing that would remain after combustion is ash. Heating coal to a high enough temperature in the absence of air so that it can't burn removes essentially all of the volatile material content and is defined by the term *coking*, and that which remains is called the *coke*. The material remaining (the coke) consists principally of fixed carbon, ash, and a small percentage of sulfur. The ash in this instance does not come from burning but is composed of stone, sand, and more (sometimes called *bone*) that was preexistent in the coal. So coke is basically a block of carbon. Sugar cokes, too; it's the black mess (carbon as well) that remains in the pot when you cook it for too long.

Since the coke in which we are interested was made from coal, it is probably not a surprising fact to find that, according to Joseph Weeks, a noted authority on the coke industry of the period,

> in 1816–'17, Colonel Isaac Meason built the first rolling mill erected west of the Allegheny mountains, to puddle iron and roll iron bars, at Plumsock, in Fayette [County], Pennsylvania. At this mill . . . coke was used in the refinery [furnace]. This is the first definite statement that I have been able to find of the use of coke in this country. It is an interesting fact that it was made on Redstone [Creek], about midway between Connellsville and Brownsville, in Fayette [County], the county that produced the largest number of tons of coke in the last census year [1880].[13]

Coke, at this time (eighteenth and nineteenth centuries), was made in *ricks*, piles or heaps, and also in *beehive ovens*.[14] Beginning in the late nineteenth and early in the twentieth century in the United States (somewhat earlier in Europe), coke was also manufactured in *retorts*, sometimes called *by-product* or *Belgian ovens*. The term *ricks* is derived from the manufacture of charcoal. Wood was heaped in piles on the ground, the piles were then covered with dirt and wet leaves, and the wood was set on fire. The covering of dirt kept the heat inside. A person familiar with the process would extinguish the fire at the proper time (after a number of days), long before all of the wood was consumed. The intense heat converted the remaining wood into a light, fluffy form of carbon that we call charcoal. This same process was later applied to a pile of bituminous coal, making coke instead of charcoal.[15]

Another invention that was important to the growth of the metals industry came from eighteenth-century England—the beehive coke oven, so named because of the earliest form of the closed kiln's resemblance to a beehive.[16] A beehive oven was a refractory-lined hemispherical chamber, typically eleven to twelve feet in diameter and six or seven feet high, with a twenty- to twenty-four-inch hole or *trunnel head* in the top or *crown* of the oven, and a doorway at the bottom on one side. The flat-bottom floor on the inside of the chamber was paved with brick. In this country, the hemisphere of brick was usually completely covered over with earth to form a level surface, except the space for the trunnel head was left open. Externally, a grouping of beehive ovens looked somewhat like a retaining wall holding back the hillside, with the exception of a hole visible in the side of the wall for each oven, along with another one on top in the earth cover (see figures 3.1 and 3.2). Three to six tons of coal were loaded through the trunnel head or charging hole; the conical-shaped pile of coal that formed inside of the oven was then leveled by a man using a hooked bar inserted through the door on the side. He used the bar to form a uniformly level bed whose depth was typically twenty-four to thirty but rarely thirty-six inches thick. The doorway was then walled up with brick and plastered closed with mud (*luted*), except for a small air passage. Either forty-eight or seventy-two hours later, based on the depth of the leveled coal (sometimes ninety-six hours for Thursday charges to avoid Sunday work), the wall was pulled down, the coke was watered in the oven to extinguish it, then it was pulled (*drawn*) from the oven with forks and rakes (see figures 3.3 and 3.4). An excellent description of the actual operation is found in Weeks:

> [at the] works of the Morewood Coke Company . . . The method of operating these ovens in the Connellsville region is quite simple, and may be taken as the usual practice in this country. The coal is generally brought to the oven in lorries holding each a full charge, 125 bushels, for 48-hour furnace coke. The lorrie is run to the charging-hole on a railroad over the top of the oven, and the coal is dumped through the hole in the crown of the roof and carefully leveled by means of a long iron hook inserted into the door. This door is bricked up and plastered up or daubed, except for some small interstices at the top, so as to admit only sufficient air above the coal to carry on combustion. The heat which the oven acquired in the preceding operation is always sufficient to ignite the new charge, combustion being carried on by the entrance of air through the doorway, and the coal soon begins to emit aqueous and sulphurous vapors, followed by a thick, black smoke and reddish flame around all sides. At this stage of the process the gasses are particularly offensive. The heat of the oven at this time is a low red. In a few hours the mass of burning coal cracks downward, enabling the volatile matter below the surface to pass off, and by its ignition to generate additional heat for carrying on the process. In about 12 hours a clear, bright flame prevails over the entire surface, which increases almost to a white heat. Basaltiform columns are formed, which allow the gasses to rise as the heat ascends. Finally the clear, bright flame dies off gradually, and the coke becomes a glowing red mass. Now the sooner the oven is quenched and drawn the better, for the coke will continue to take up air in spite of every precaution, and the red-hot coke will waste, lose heat, and become an inferior fuel.[17]

A bit more insight is gained from a further passage about Durham, England, Connellsville's chief competitor for the title of King of Coke, where it is described as follows:

> The coal is fired at the surface by the radiated heat from the roof, enough air being admitted to consume the gasses given off by the coal, and thus a high temperature is maintained in the roof of the oven. The coal is by this means melted, and those portions of it which, under the influence of a high temperature, can of themselves form gaseous compounds.[18]

Although the process of making coke is referred to as *burning*, that is very misleading. Only enough air to burn the gas collected in the crown of the oven above the coal was permitted to enter the oven. Any air above that amount, termed *excess air*, would cause the coal/coke to burn and be wasted. Because the air was controlled in this way, there was not enough for complete combustion, and the gas emitted from the port in the top of the oven tended to be very smoky, especially during the early part of the coking cycle. Coal in coke ovens is actually *distilled* and not burned; that is, heated in the absence of air to vaporize the volatile material.

Employment at a beehive coke plant tended to be long, hot, dirty, difficult work performed under poor conditions for little pay. It required few skills and no education, just a strong back. "Americans" (at that time men who were of English, Scottish, Irish, German, or Austrian descent) often tended to band into organized groups and form unions; consequently, mine and coke plant owners would import immigrants, primarily from eastern European countries, to do this work. The "Americans" frequently referred to those people of eastern European ancestry as *Huns* or *hunkies*, this berating and somewhat contradictory term reflecting their "foreign" origins. Sometimes black men were brought in from the southern states to discourage unionization efforts.

The pay for each job was usually determined by the task, not at an hourly or daily rate. A few examples of oven workers' pay scales from 1903 are as follows: drawing one oven (quenching and removing about three tons of coke), $1; leveling an oven, $0.09; charging an oven, $0.025; forking cars of forty thousand to sixty thousand pounds (loading railroad cars with coke), $1.50 to $1.60 per car.[19] From data compiled during the 1894 to 1900 era: drawing coke (including pulling down the wall and watering) per one hundred bushels of coal charged, $0.43 to $0.72 (note: a bushel of coal was approximately equal to eighty pounds, therefore one hundred bushels of coal represented an amount of 2.25 to 2.5 tons of coke), and leveling, per oven, $0.075 to $0.12.[20]

Connellsville coke, made from Pittsburgh Seam coal in the Connellsville basin, was a *benchmark* coke, and it was probably *the* benchmark against which all other cokes in the world were measured, including Durham coke.[21] Routinely it was made in beehive ovens located at the mine, usually from coal known as *run-of-mine*[22] or even *slack* coal.[23] Run-of-mine usually refers to coal that was *as mined*; that is, there was no processing done to the coal. It was not screened or washed; it was just as it came out the mine in cars. Coal was usually sold as *lump* or *nut*, which indicated the size of the product. Slack was that part of mined coal that was smaller than the nut coal and was considered waste to be disposed of on the *slag* pile. Some Pittsburgh ovens were charged with 100 percent slack (sometimes also called *slag*) coal.[24] The amount of slack coal generated was considerable. According to the Geological Survey of 1886, "Every square mile of a horizontal coal-bed may be said to yield a million tons to every foot of coal-bed—that is, for a ten-foot bed, 10,000,000 tons; or allowing for one-half waste, 5,000,000 tons."[25] This does not mean that 50 percent of the coal that was mined was slack. The larger portion of that percentage left behind served as pillars to support the mine roof. Other coal was too dangerous or difficult to remove, and some was too hard to separate from the waste. Still, much of this was slack. Therefore, Connellsville coke, arguably the best coke in the world, was often made by using a considerable portion of what others would call waste. The implications of this are enormous, both in terms of yield and cost of production. This extended the life of a valuable natural resource while simultaneously reducing some of the environmental implications, not that the environment was usually taken into consideration. In addition, miners were frequently not paid for mining slack coal that was considered waste, a recurring point of conflict between miners and mine owners.

The first coke ovens in the Connellsville region, or in all of America for that matter, were two built in 1841 by a couple of carpenters on the farm of a stone mason, John Taylor, who was mining coal in a "small way." The carpenters, McCormick and Campbell, built two barges. They hauled them laden with the coke down the Youghiogheny River, where the farm was located, to its junction with the Monongahela River and thence down the Ohio to Cincinnati, where they sold all of the coke after much difficulty. Before that time, Connellsville coke was made "on the ground." Their enterprise was deemed a failure, and they made no more coke.[26] A year later, foundry men from Cincinnati, who used the material and recognized the value of their enterprise, came to the Connellsville region in search of more, but McCormick and Campbell refused.

By 1855 there were 106 ovens in the United States, all on the Pittsburgh Seam. In 1875 it was 3,826, again all Pittsburgh as well. Not until 1880 do we find ovens listed in other areas of the United States, and even then, it was only a quarter of the more than twelve thousand in existence.[27] The value of this product was widely recognized. At its peak in 1910 (using total oven units as a measure), the number of beehive coke ovens in the United States was 100,362, of which 54,360 were in Pennsylvania, which included about

40,000 in the Connellsville District or 40 percent of the U.S. total. During 1916, the peak beehive coke production year in America (as determined by tons of output), 35,464,224 tons of coke, were made. Some 27,158,438 of these tons came from Pennsylvania, with 21,654,502 attributed to Connellsville, an amazing 61 percent of the nation's total.[28] Almost all was used in the manufacture of iron.

From 1916 forward, beehive production began to decline, and beginning in 1919 more coke was made in long, narrow (roughly thirty to fifty feet long, about seven to ten feet high, and about seventeen to thirty inches wide) refractory retorts frequently called by-product or slot ovens, because the gas produced was extracted, and chemicals, typically coal tar, ammonia, and light oil (which in addition to the gas were the by-products of coking, hence their name), were recovered. A portion of the gas was returned to external flues where it was burned to maintain the heat for the coking process to continue. The excess gas was used at the plant for making steam or for heating other types of furnaces. This gas, as well as all of the chemicals, was wasted (burned) on beehive ovens. World War I was the driving force for this change away from beehive production in the Pittsburgh area, and not "modern" coking methods. Ammonia and light oil were some of the by-products of coking. Light oil was composed mainly of three ingredients—benzene, toluene, and xylene—which could be easily refined into these pure products. The Germans dominated the by-product coking industry of this time. Toluene was an important component of a compound called trinitrotoluene that was needed for the Allied war effort. Its more familiar name is TNT. Ammonia was also used for explosives.

U.S. Steel built the largest by-product coke plant in the world at Clairton, Pennsylvania. It went into production in 1918 just as the rest of the nation had been gearing up war production of these important chemicals. Another large plant, built by Jones and Laughlin Steel in Pittsburgh, started operating shortly after the war ended. Good-quality coke could be manufactured with lower-quality coals that were crushed, washed, and blended for use in these "new" type ovens. Excess gas could be used elsewhere, lowering costs. Unlike beehive ovens, these by-product ovens could not be easily shut down for a number of reasons. They might sustain extensive damage to the refractory that was difficult and expensive to repair. So despite the fact that by-product ovens were thirty to forty times the capital cost of an equivalent-producing beehive plant, they were newer facilities at the steel plant, and needed to be run continuously to be effective. There were other benefits as well. Coke yield per ton of coal was about 7 to 10 percent better than a beehive oven, and the chemical sales helped to offset production costs. Beehive plants were shut down and abandoned in accelerating numbers due to a postwar slowdown in the steel business, as coke requirements warranted, and more of these newer plants were built. Using blends of lower-quality coals in the new ovens vastly increased the number of regions where metallurgical coal could be obtained, and consequently the Connellsville region never again held the mystical power that it once wielded over the American steel industry.

PRODUCER GAS

As previously mentioned, many puddling and crucible furnaces were heated through the use of a fuel called producer gas, and this would also include open hearth furnaces. Later, during the era of mass-produced steel, the use of gas simplified operations at the furnace, especially when considering coke-fired crucible furnaces. Producer gas was made from coal (or coke) in a simple process akin to coking. "Producer gas has been among the most popular types of gaseous fuel since it can be made in a continuous process, requiring little supervision, from low-grade solid fuels."[29] The coal was converted into gas in a cylindrical (or rectangular or pyramidal) bed about five feet in depth. The bottom foot or so of the bed was ignited. This was where air and frequently steam (or water to make steam) was injected. Steam was usually used as the means of propelling the air into the producer, but it also participated in the reaction to make the gas. Coal was added at the top and slowly descended through the vessel. A chemical reaction occurred near the bottom caused by the decomposition of the steam and its combination with the air and carbon of the incandescent coal. The gas generated was composed primarily of carbon monoxide and hydrogen. It had a heating value in the range of 110 to 160 British thermal units (BTUs) per cubic foot, or somewhere between a sixth and a tenth of that of natural gas. A hand-fired machine could be expected to gasify six hundred pounds of coal per hour to provide sixty-five cubic feet of 140 BTU gas per pound of coal.[30] Pittsburgh coals were an excellent raw material for the production of producer and other forms of manufactured gas due to their volatile matter content being in the 30 percent to 40 percent range.

NATURAL GAS AND OIL

In 1878 a gas well was drilled near Murrysville, Pennsylvania, about eighteen miles east of Pittsburgh. Although not the first well drilled in America, it was considered to be the first *commercial* gas well in the country in addition to being one of the most productive.[31] The gas from the Haymaker Well, as it became known, was eventually piped to Pittsburgh and its environs and was used in iron and steel mills, glass factories, other factories, and hotels. By 1886 there were already sixty-six miles of gas lines within the city, some as large as twenty-four inches in diameter. An estimate made by a representative of the gas company, published in the *American Manufacturer* of August 13 of that year, determined that the *daily* displacement of coal by natural gas in the city of Pittsburgh alone was twenty thousand tons.[32] Considering the source, the estimate is probably high, but even so, it was still an incredibly large figure. Gas was found in other areas near Pittsburgh, "and pipes from one of these wells was laid a distance of some 17 miles to the iron works of Spang, Chalfant & Co. [in Etna], the Isabella Furnace and the iron mills of Graff, Bennett & Co., in all of which the gas is used as a fuel in the place of coal."[33] Ironically, Spang was one of the early producers of seamless pipe in America.

Not that it had much bearing on the development of the steel industry, but more to understand the totality with which Pittsburgh abounded in fuel: in August 1859 a man by the name of Edwin Drake drilled what many have called a "quite unremarkable hole" just to the south of Titusville, Pennsylvania, about eighty miles north of Pittsburgh. This unremarkable hole turned out to be the world's first purpose-drilled oil well. Before that time, wells drilled near the same location in search of drinking water or brine for salt production (an important regional industry that also consumed a great deal of fuel) often had to be abandoned because of the "useless" gooey mess and the bad taste imparted to it by the oil. Drake's well produced only about twenty barrels of oil a day. Still, as soon as it became known, others quickly began drilling in the area, and a new industry was born. Until that time whale oil or tallow was used for lighting homes; afterward kerosene used for this purpose was abundant.

In 1854 Samuel Kier established the first petroleum refinery in America, located in Pittsburgh. Earlier small-scale refining experiments were conducted elsewhere. Oddly, the Pittsburgh refinery predated Drake's well by five years. Kier's *carbon oil* or kerosene, together with the new lamp he invented to burn it, transformed the world with its brilliant light. For a considerable time, oil had been collected in the Oil City-Titusville area, where it was found floating on lakes and ponds where it had gathered from seeps in the ground or in brine or water wells. Kier collected the material from his father's brine wells and frequently sold it as patent medicines. He could not dispose of all of the oil that way, so he started to experiment with refining. Since pipelines were virtually nonexistent, much of the oil was shipped by barge to Pittsburgh on the Allegheny River until the industry developed and lines were laid. One estimate indicated that as much as 50 percent of the oil shipped was lost before it reached the city due to leaking barrels and river accidents. Samuel Kier was also a partner in the firm that was the forerunner of Jones and Laughlin Steel.

TRANSPORTATION

The city of Pittsburgh grew at the headwaters of the Ohio River, where it is formed by the confluence of the Allegheny River, flowing in from the north, and the Monongahela, more commonly called the "Mon," whose origins are to the south in West Virginia. There were two other navigable tributaries that played a role in the region's river network: the Conemaugh-Kiskiminetas River system that entered the Allegheny at Freeport, and the Youghiogheny River and its connection with the Mon at McKeesport. Pittsburgh was the original Gateway to the West. This role was enhanced with the completion of the Erie Canal in 1825. Although the city is about 180 miles as the crow flies from the Erie Canal's terminus in Buffalo, it was a short, relatively level overland jaunt to the Allegheny River in northern Pennsylvania, thence through to Pittsburgh by river, and finally as far south as the Port of New Orleans or as far west as can be reached from the Missouri and its tributaries. The principal alternate route eastward to Philadelphia or New York City was by an almost one-hundred-mile long, mostly uphill climb over the Allegheny escarpment and then from there approximately 250 more miles via a series of rivers and roads.

Just as Pittsburgh's rivers played a unique role in exposing the region's vast mineral resources for economic development, there also existed a symbiotic relationship between these rivers and those exploited assets that fueled the engine of industry in their vicinity. Due to the large volumes and immense tonnages involved, these arteries were a critical link for both the distribution of raw materials and dispersal of finished goods. When it's available, water transportation is generally the lowest-cost shipping mode for large-volume or high-tonnage materials. The Allegheny Valley was heavily populated with industries that might be termed "little steel," such as wrought iron and crucible steel manufacturers, or the smaller, but certainly extremely important, early plants. It served as the breeding ground for the later massive growth that occurred along the Mon during the mass-produced iron and steelmaking era. To give you some idea of what is meant by massive, Pittsburgh was probably the most heavily industrialized region ever to exist on the face of the earth. Substantiating such an assertion is difficult to say the least, but quoting William Wilson, who authored a history of the Pennsylvania Railroad: "It may be a surprising statement, and yet is true, that in 1893 the tonnage by river and rail from the Monongahela Valley exceeded the entire Trans-Atlantic tonnage of the United States, and also that to South America."[34] To demonstrate that Wilson's statement was not just an aberration, figures published in *Pittsburgh First* of March 6, 1920, the Official Bulletin of the Chamber of Commerce, reported that 1918 Monongahela River tonnage was sixteen million tons. This exceeded the combined tonnages of both the Panama and Suez canals that were only 7,562,000 and 7,833,000 tons, respectively.[35] Similar figures are available for 1919 and 1945 and likely other years as well.

While the marriage between heavy industry and the local river network might at first appear to be a perfect union, especially in the case of the Mon, the rivers at Pittsburgh were not capable of providing a year-round *slackwater* navigation environment. *Slackwater* refers to river water levels that were deep enough and slow enough to provide an unobstructed path for barges and riverboat traffic. This was not the case for the Pittsburgh harbor many times each year, especially in late summer and early fall. During these times the rivers could be waded or walked across at many locations. In spring, snowmelt and rain would sometimes make the rivers run too fast and treacherous to use, and by winter they would often freeze. This was one of the reasons people thought that Brownsville, about forty miles south on the Mon, or Wheeling, (West) Virginia, on the Ohio, would become more important than Pittsburgh. Both were located on the National Road (now US Route 40), and both were river ports. In addition to being the original terminus of the National Road in 1820, Wheeling was the easternmost port on the Ohio System to offer year-round slackwater navigation.[36] Later, the Baltimore & Ohio Railroad began serving Wheeling, connecting it to the Chesapeake Bay, thereby increasing its importance. To counter these deficiencies, Pittsburghers encouraged the federal government to develop a series of locks and dams to promote navigation in the region, but to no avail. The same problems were encountered when seeking aid from the state of Pennsylvania. A recurring theme in Pittsburgh's rise as an industrial and financial center was the never-give-up attitude of its entrepreneurs. Failing to get federal involvement, local people formed a capital stock company called the Monongahela Navigation Company to build, maintain, and operate a series of locks and dams on the Monongahela River. By 1841 the first two locks and dams were put into operation, and at each dam the river was raised eight feet.[37] By 1844 locks and dams were built near Elizabeth and Brownsville, and somewhat later others provided for navigation as far south as the (West) Virgina state line. There were seven in all. Elizabeth (or that general area) is important for other reasons as it was the place where the boat for the Lewis and Clark expedition was made and was the location from which one of them departed on his famous adventure.

Another advancement that addressed the dilemma of low river water levels was the development of "The Pittsburgh steamboat . . . a specially constructed craft [with flat bottoms and shallow draft], designed to navigate these comparatively narrow and shallow rivers. The transport service or 'towing' of these steamers is quite unique, and as a consequence, Pittsburgh counts steamboat-building a feature of her many industries."[38] In 1811 the

New Orleans, the first steamboat to ply the western waters, was built in Pittsburgh. "Voyages of 4,000 miles are quite common, and the farthest recorded trip of a Pittsburgh steamer is to Cow Island, on the Upper Missouri, 4300 miles distant."[39]

One might think the contention that these astonishing statistics were primarily attributed to only one industry would be a gross overstatement, yet the prevailing factor was overwhelmingly the steel industry. This hypothesis is substantiated by the following facts: In 1921 the Monongahela River transported 26,374,682 tons of freight, of which roughly 21,000,000 tons was coal, 2,600,000 tons of sand and gravel, 750,000 tons of coke, 500,000 tons of steel, and the remainder all others. On the Allegheny the total traffic was 3,761,739 tons, which included 2,300,000 of sand and gravel and 1,200,000 of coal. While for the Ohio, as far as Beach Bottom, West Virginia, the figures were 10,158,478 tons with 5,400,000 being coal and 3,700,000 sand.[40] The 8.6 million tons of sand and gravel, admittedly a large portion of the total, is not indicative of heavy manufacturing industry and is in part due to the need to maintain the depth of the river channel. Discounting all other shipments except those related to coal and steel (a minor fraction) breaks down to 22,250,000 tons for the Mon, 5,400,000 on the Ohio, and 1,200,000 carried on the Allegheny. Even by going as far as assuming that no coal on the Ohio or Allegheny was mined there and that all of the tonnage accounted for on these two rivers was exported there from the Monongahela, that would still represent slightly less than sixteen million tons not leaving the Monongahela Pool. While some of these coal tons were surely used for heating or other purposes, the bulk of this river's tonnage was intimately associated with steel, whose major steel plants were confined to a roughly twenty-five-mile stretch of the river.

The Pennsylvania Railroad was the largest in the United States and possibly the entire world, at least when evaluated on a tonnage basis. For some time it was the largest corporation in America. Pittsburgh was the freight traffic center of this railroad. As you can see from table 3.2, the total number of cars in and out of Pittsburgh for the month of August 1917 was 277,210. The Pennsylvania Railroad's share was 64,350 cars or roughly 25 percent of the total. Two other railroads seem to stand out, not because they are large, but rather small, in addition to being virtually unknown by most people: The Monongahela Connecting RR (MCRR) was a small intraplant railroad owned by the Jones and Laughlin Steel Company (J&L) in Pittsburgh. It was only about five miles long. The Union RR (URR) was an interplant switching railroad formerly belonging to the Carnegie Steel Company, but at the time of this analysis, it belonged to the U.S. Steel Co. (USS). The URR connected that company's plants in Homestead, Rankin, Braddock, Duquesne, and

Clairton, Pennsylvania, with the rail yard of the Pittsburgh, Bessemer & Lake Erie RR (later B&LE) at North Bessemer, Pennsylvania. The B&LE was a USS (Carnegie) owned road as well. The Union operated for a distance of approximately twenty miles along the Monongahela River, interconnecting its plants in that valley with a line about nine miles long that went up through the Turtle Creek Valley to the north where it had a junction with the Bessemer. The primary function of the Bessemer was to bring iron ore down from the lakes and return coal or coke to them. Yet if you exclude the Monongahela connecting's contribution to the total, the Union's sum of 109,800 almost equals the combined car total of 122,360 for all other railroads in Pittsburgh! So the combined total of two small steel industry railroads located along the Monongahela River, which were about thirty-five miles in total length, far exceeded the sum total loadings and unloadings of cars of all other railroads in Pittsburgh together, including that of the largest railroad in the United States, whose system had in excess of twenty thousand miles of trackage. One must understand that these numbers do not reflect the size of through traffic (cars on trains that were passing through; that is, that did not originate or terminate in the region) on those roads, or the fact that one or more of them had to deliver or accept delivery to ship the cars to and from the URR or the MCRR. Those numbers were not added to their total. Also, some of the loads for the Bessemer (counted as URR) were likely coal for export to other USS plants outside of the Pittsburgh region. On the other hand, many cars on the West Side Belt and Pittsburgh & Lake Erie (P&LE) were associated with the Monessen Works of Pittsburgh Steel, as were the PRR's share from the Donora plant that were served by those rail lines, while the National Tube Works was serviced by the P&LE and the B&O. Additionally, although the information is a bit outside of the timeline for this work, taking into account numbers compiled when Pittsburgh was on its "downward swing" as a steel center, the Borough of Port Perry, across Turtle Creek from the Edgar Thomson Works of USS in Braddock, had the highest freight traffic density in the entire world. At this location the Pennsylvania Railroad's Port Perry Branch united its Pittsburgh and Monongahela Divisions, and the Union RR main line crossed over the Monongahela River on parallel bridges. Both are situated slightly upstream of Lock and Dam No. 2 on the Mon. These bridges crossed over the parallel main lines of the Baltimore & Ohio and the Pittsburgh & Lake Erie railroads near the same point in Port Perry. This grouping of important transportation facilities is the reason it was given the esteemed title.[41] All of this rail and river traffic converging at this location attests to the importance of the area.

The Mon is one of those rare rivers that flow from south to north. The steel industry in Pittsburgh was concentrated at the northern end of the Monongа-

Table 3.2. Freight carloads in and out of the Pittsburgh district, August 1917. Cars were only counted once, at the point of delivery or shipment, and are not duplicated between railroads. Compiled from *Railroads: A Part of the Pittsburgh Plan*, 1924.

Railroad	*Carloads In*	*Carloads Out*
Pennsylvania (PRR)	35,000	29,350
Baltimore & Ohio (B&O)	7,870	6,330
Pittsburgh & Lake Erie (P&LE)	12,720	12,450
Pittsburgh & West Virginia (P&WV)	620	320
West Side Belt	3,530	6,720
Pittsburgh, Chartiers & Youghiogheny	1,780	4,050
Montour	100	1,520
Monongahela Connecting (MCRR)	20,450	24,000
Union including Bessemer & Lake Erie (URR and B&LE)	56,800	53,000
Total	138,870	137,740

Source: Municipal Planning Association, *Railroads of the Pittsburgh District, a Part of the Pittsburgh Plan*, Report No. 5, Citizens Committee on City Plan of Pittsburgh, 1924, 17.

hela Valley near the city. Coal mines flanked the valley along its length, but coking was clustered along the tributaries to the south. So a synergy existed along this little-known river, between the heavy industry concentrated at the northern end and coal and coke manufacturing on the other. This microcosm on the Mon, a network of steel plants confined only to some tens of square miles connected by river or rail to the coal and coke lands in the south whose total area was only a few hundred square miles (smaller than the land area of some modern American cities), had a total output that favorably compared with or exceeded that of the world's major industrialized nations!

There was a time when one could take a boat ride on the Mon and never be out of sight of a major iron or steel facility for roughly the first seventeen miles. This did not include any of the smaller iron and steel manufacturers found along the way, of which there were many. From a modern perspective it seems somewhat amazing that the very first plant that would have been encountered was the Clinton Blast Furnace of Graff Bennett Company that was located directly across the river from downtown where one would have boarded. From there the boat would have passed the J&L plant located on the south side of the city and as well as in Hazelwood; the Mesta Machine Company in West Homestead; the Homestead Steel Works (USS) in Homestead, Munhall, and Rankin; the Edgar Thomson Steel Works (USS) in Braddock; the Duquesne Steel Works (USS) in Duquesne; and finally the National Tube Company (USS) works in McKeesport. Yet in these seventeen miles, one would not have seen the Clairton or Donora Works of USS, nor the Monessen Works of the Pittsburgh Steel Company that were located farther along. At one point in time on the river trip to Monessen, the world's largest beehive coke plant in Hazelwood and the world's largest by-product coke plant at Clairton would have been passed. Upon reaching Monessen one would have been well within the beehive coking boundaries of the Pittsburgh region. Remember, of the fifty-four thousand ovens in Pennsylvania, the greatest concentration in America, about forty thousand were located near this town in the Connellsville coking district. During the day the vast amount of steam released from quenching the incandescent coke tended to combine and intermingle with the smoke streaming from the oven ports, and would often in the coolness of twilight form into strings of mist along the valley floors. In the dim evening light, the red, yellow, and blue flames produced by these gasses venting from each of the beehive ovens were perceived by many in a romantic sense to be little jewels imbedded in the mist. This impression, repeated thousands and thousands and tens of thousands of times in valley after valley on both sides of the Monongahela River became known as the "necklace around the throat of industry" in Pittsburgh.

Figure 3.1. Cross section of a typical beehive oven in the Connellsville Region.
Modern Beehive Coke Oven Practice, RI 3738, U.S. Bureau of Mines, December 1943.

Figure 3.2. A bank of beehive ovens in the Pittsburgh region. This type oven was very smoky due to the fact that air for combustion inside the oven chamber was limited in order to prevent burning of the coal or coke. This created a smoky flame in the expelled gasses, especially during the early part of the coking cycle.

Photo courtesy of Rivers of Steel National Heritage Area.

Figure 3.3. Coal is being charged into an oven (top) through the trunnel head using an electrically driven charging car or larry, probably a corruption of the British lorrie for truck. Below, a worker loads a wheelbarrow with coke that was just removed from the oven.

Photo courtesy of Rivers of Steel National Heritage Area.

Figure 3.4. Men "drawing coke," that is, removing coke from beehive ovens. Three ovens per day per man was typical or about 9 tons of coke a day. Other men (above) are waiting to refill the ovens with coal. This was hot, dirty, difficult work.

Photo courtesy of Rivers of Steel National Heritage Area.

Chapter Four

Iron

Iron is one of the most important materials known to man, so significant in fact that it lends its name to one of the three ancient Ages of Man, the Iron Age. (The Stone Age and Bronze Age are the others.) The Iron Age lasted for well over a thousand years, until the modern age of man, and is generally accepted by those of European heritage to mean that it ended in the Greek and Roman era. Present-day society is still deeply dependent on this metal, but ours is the steel age.

The subject material was and continues to be formed through nucleosynthesis in the hearts of massive stars. Stars are energized through thermonuclear fusion of hydrogen, helium, and more—the primary components that comprise a star. The fusion process is ignited by gravity, which caused the material to collect in the first place. Stars normally "burn" or fuse heavier and heavier elements, as the lighter and easier-to-fuse hydrogen and helium become depleted. For stars that are about one-and-one-half or more times the mass of our sun, the nuclear "burning" is more rapid, and for these stars, the higher the mass, in a somewhat counterintuitive sense, the shorter the life of the star. The more massive portion of stars (about ten times the mass of our sun or more) can "burn" elements in successive steps until a stable form of iron is formed. For elements heavier than iron, more energy is required for fusion than is released. So iron represents some form of stability. While the iron that has just been synthesized is stable, the star is not. Because there is no longer enough outward energy from fusion to counterbalance the inward force of gravity, the star's atmosphere rapidly collapses onto its core and causes an explosion. We call this explosion a supernova, and it briefly outshines the entire galaxy, comprised of billions of stars. These gigantic detonations disperse the iron from the cores of the stars to the netherworlds of the universe, including our Earth.

Iron is one of the most common materials found on Earth. Except for *meteoric iron* or iron that has crashed into the Earth from outer space, it is rarely found in nature in its pure form. Our planet's iron is usually found as an ore, which is chemically bound with oxygen or a similar element. The secret of mastering this material is in breaking those bonds. This involves much more than just melting the ore. So the challenge for man is to overcome and break the bond between the iron and the oxygen in the ore (iron oxide), the form in which it is most commonly mined. This process is like those myriad of things that are simple and easy to understand but difficult to put into practice.

Surely, some ancient man first recognized and then understood that certain rocks that were subjected to a windblown wood fire left behind small flecks of a hard, shiny material that we know as iron, and then used this knowledge to his advantage. Eventually this led to understanding that the charcoal from the wood and the fire, driven to high temperature by the wind, was key to freeing the iron from this peculiar rock. In Britain, the Romans built furnaces that were called "*bloomaris*—a simple open furnace with a blast produced sometimes artificially, and sometimes by leading a draught [air] passage out from under the hearth toward the side from which came the prevailing strong winds."[1] Britain was being denuded of trees by the time of the American colonization. Consequently, it looked to the new world as a potential source of iron because of the large forests there that could be converted into charcoal for smelting. From 1558 through 1584, during the reign of Elizabeth I, various acts of Parliament were passed with regard to the permissible location of ironmaking furnaces due to the scarcity of wood for charcoal.[2]

To understand how much wooded land was actually required to feed a smelting operation, an acre of forest could be expected to yield about fifty cords of wood.[3] Citing Overman, about forty bushels (twenty-five to forty-five bushel range) of charcoal could be produced from a cord of wood (although these cords may have been of different sizes).[4] He also notes that, "3 tons of ore and 200 bushels of coal [charcoal] are required for 1 ton of iron,"[5] or the equivalent output of about ten tons of iron per acre of wood. So a furnace producing fifty tons per week (a reasonable figure) would require about 260 acres of forest per year. According to Malcolm Keir,

> So long as furnaces were small, charcoal was the most desirable fuel because it made the purist iron, but a large furnace required so much fuel that the surrounding wood was soon exhausted. . . . Thus, if all the iron of 1850 [made in the U.S.], for example, had been produced by charcoal, it would have consumed 46,000 acres of forest. On the other hand, 200 acres of four-foot coal [anthracite] could do the same work. A moderate-sized furnace necessitated a wood reserve of 4 square miles.[6]

Investigating the chemistry of the process revealed that the carbon of the charcoal provided a means of removing (known as *reducing*) the oxygen bonded to the iron. When carbon monoxide was formed during the combustion of the charcoal, the iron could then be freed from the ore through the reducing reaction in which the carbon monoxide combined with the oxygen in the ore to form carbon dioxide. In breaking the bond of the iron oxide, the ore became elemental iron. The charcoal also served as the fuel to provide the energy (high temperature) to break the bond. Then, the free carbon acted as an alloy with the solidified iron, giving the material its hardness.

Early furnaces were short shafts only a few feet high. See Percy for a brief description of the operation in India of these small-shaft furnaces that produced from a few pounds to a few hundred pounds of iron at a time.[7] No liquid iron was cast. It was removed as a solid mass mixed with cinder. Later, charcoal blast furnaces looked somewhat like truncated stone pyramids that had central, open shafts where the iron was smelted. Some looked like truncated cones. Typically in the eighteenth and nineteenth centuries they were about forty feet at the base and forty feet in height, although the numbers varied greatly in both dimensions. Frequently there was a bridge or ramp at the top level or an elevator with a ramp to facilitate the loading of raw materials. Charcoal, iron ore, and limestone were added through the top of the open shaft (see figures 4.1 and 4.2). A blast of cold air at ambient temperature

Figure 4.1. An example of a stone construction charcoal era blast furnace with an elevator for hoisting raw materials to feed the furnace.

From Metallurgy, Overman, 1855.

Figure 4.2. An example of a stone construction charcoal era blast furnace with a ramp for feeding the furnace.
From Metallurgy, Overman, 1855.

(but much later it was heated) at about one-half to one-and-one-half pounds of pressure was blown through nozzles (typically numbering two or three in early charcoal furnaces) called *tuyeres* (*twyres* in England) that were located near the bottom where the molten iron was extracted. Typical production ran from a few tons to as many as fifty tons a day. Because of the pyramidal shape of the early stone furnaces, a small increase in the height of the central shaft required a huge increase in the mass of stone required for construction due to the base width-to-height correlation. In order to surmount this shortcoming, the furnaces were bound with tie rods and/or bands of wrought iron to hold them together, thereby reducing the need for heavy stone abutment. This caused the furnace appearance to be more pillarlike.

At the beginning of the eighteenth century, about 1715 to 1740, an important shift in the process of iron smelting occurred at Coalbrookdale, England. Abraham Darby and his son, also named Abraham Darby, were the first to be credited with manufacturing iron by using coke as the sole source of fuel in a blast furnace. The Darbys learned to make coke from bituminous coal in pits on the ground. This skill was then used at the blast furnace, where additional experimentation, including increasing the blast pressure, was required for success.[8] The motivation, of course, was the elimination of reliance on scarce charcoal. Others had attempted to use raw coal and/or coke to wholly or partially replace the wood-based fuel but had not met with success. In America these innovations were not adopted satisfactorily for about another one hundred years, until 1837 to 1840, when coke alone was used to fuel first the Lonaconing and later the Mt. Savage blast furnaces in Maryland.[9] As part of this development, coke furnaces evolved into the metal-encased, refractory-lined vessel with a shape that would be recognizable as a relative of present-day furnaces. By using plates and hoops as a binding, additional excess material could be removed at the corners, and the vessel took on the appearance of a

Figure 4.3. A cut-away and cross section view of a blast furnace at Hyanges, France. Note that the external shape is maintained by metal plates covering a stone and refractory lining held together by metal hoops.

From Metallurgy, Overman, 1855.

cone or cylinder. By eliminating more material at the *hearth* (where the molten iron was collected) and by supporting the shaft of the furnace on piers in place of the removed stone, the furnace took on a more modern appearance, as shown in figure 4.3 at an installation in Hyanges, France.[10]

The internal shape of the furnace shaft evolved as well. For a comparison view of some early internal shapes, see figure 4.4. The shaft is frequently described as a series of interconnected conic sections, whose narrowest

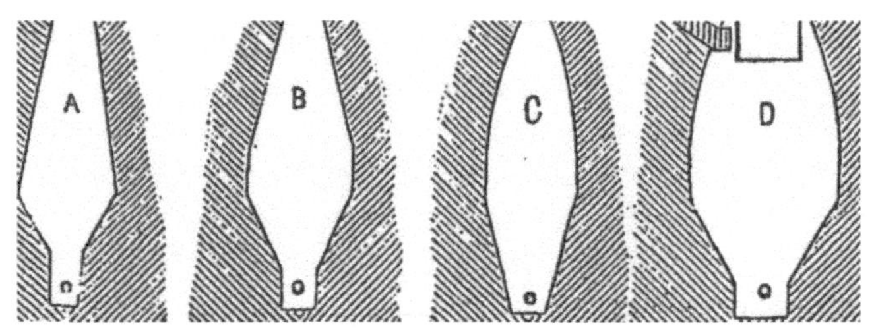

Figure 4.4. Variations in the internal shapes of the shafts on early furnaces.

From Metallurgy, Overman, 1855.

dimension is at the top, continually enlarging as it descends. Near the bottom it began to shrink again, almost like an elongated raindrop that is nearer the elliptical shape of some early furnaces. The uppermost part is called the *stack*; the widest section the *bosh*; and the lowest rectangle (early) or cylinder (later) is known as the *crucible* or *hearth*. This profile is not arbitrary. It was developed through operating experience and is related in part to the descent of raw materials, as well as to the upward flow of gasses through the furnace. If the shaft of the furnace had been in the form of a cylinder, the descending material would key into irregularities in the walls that had developed through either material deposition and/or wear and then jam to form a bridge across the entire opening. This tendency to bridge was buoyed by the upward flow of the large volume of gasses passing through the shaft, which tended to lend support to the jam. Furnace operators and managers called this a *scaffold.* Occasionally the furnace had to be shut down and men had to go inside to poke at the jam with long rods to break it down, a very dangerous job. One plant in England reportedly kept a cannon on hand for that very purpose. So as the raw material passed downward the walls diverged and became farther and farther apart in an effort to eliminate this tendency to jam, until it reached the bosh. Increasing the stack width did not always accomplish the intended result. This was more true when using anthracite or coke as a fuel. Near the widest cross-section or bosh, the fuel, ore, and limestone were being consumed as the molten iron and slag were collecting just below in the area of the smaller hearth. Since the volume of the raw materials decreased, the volume requirement for that part of the furnace shaft was decreased.[11] While a major contributor, it should be stressed that this was not the only phenomenon that affected the evolution of the shape of the shaft. With regard to the narrowness of the top of the furnace, if the charging hole were too wide, the distribution of raw materials could be more inconsistent. Improperly distributed material increased the probability of channeling, which lessened contact with the flow of reducing gasses. If the top were too small, it resisted the flow of gas. By widening the furnace toward the bosh, the rate of descent of the raw material was slowed, permitting increased contact time with the reducing gasses.

One of the more important breakthroughs in the evolution of ironmaking occurred in 1828, when James Neilson of Scotland invented and later patented the hot blast.[12] Neilson's invention was a simple one that went contrary to the belief that cold air was better. This misconception was based on the fact that on average more iron had been produced in the winter than in summer. Later it was attributed to the fact that there is more moisture in summer air. Neilson simply passed the metal blast piping through a chamber that was heated with coal. The fire heated the piping, which in turn heated the blast air passing through. According to Percy, "Few persons would now demur to the statement that the hot-blast has greatly cheapened the production of iron, and in so far is to be regarded as one of the most important improvements ever made in metallurgy."[13] When the hot blast was introduced at the Clyde Works, a calculation of the fuel saved proved to be remarkable. The figures given were based on converting coal tons used to the coke equivalent: with a cold blast alone, the fuel usage was eight tons, 125 pounds per ton of cast iron; with a hot blast at 300°F it was five tons, 325 pounds per ton of cast iron; with a hot blast at 600°F, only two tons, 525 pounds per ton of cast iron were needed. (The furnace was using raw coal instead of coke for fuel at this time.)[14] A different estimate indicated a savings of 20 percent to 33 percent with a hot blast when using charcoal. There also was an increase in iron production at the furnace.[15] From another source, "The apparently singular effect of hot air in a smelting furnace . . . [indicated] a considerable amount of fuel is saved by it, which at charcoal furnaces amounts to ¼; at coke furnaces to ⅓ and at anthracite furnaces to nearly ⅗ of that used by a cold blast."[16]

Using coal to heat the air was not problem-free. The high temperature to which the pipe had to be subjected in order to transfer sufficient heat to the air, combined with the fact that there was a high differential temperature across the length of the pipe due to the air on the inside of the pipe being cold at the entry end and hot at the exit, caused high stresses within the pipes and connections, and this made them prone to leaks and failure. Others tried using the escaping waste gasses exiting from the top of the furnace stack. Contrary to what one might think, the top gas from a furnace is not especially hot, 400° to 450°F, while at the hearth some fifty-sixty-seventy feet below, the temperature was 3,000 to 4,000°F. Much of the energy from the inferno below was expended on its journey upward through the process of driving off water from and heating the vast amount of raw material contained within the shaft. Also, it had begun the reducing reactions on the ore descending through the stack.

In the later part of the nineteenth century, the addition of cooling plates to the outside of the furnace stack, which were inserted between the rows of bricks, had an effect, too. Present-day furnace tops operate in about the 200 to 300°F range, so there was not that much energy readily available unless the system used an extensive piping network that would have interfered with charging operations. Some older open-top furnaces had chimneys to vent the flames away from the men charging the furnace. While this may seem contradictory to previous statements, the combustion did not occur until the gas exited the stack and mixed with sufficient air. There were systems devised and used in some places to take advantage of that fact, but apparently, they were not very successful. Many believed that this gas could be used in place of coal to heat the blast air, and they devised ways of using it. This led to the next major evolutionary change to the blast furnace.

In order to divert the gas exiting the furnace for use as a replacement fuel for coal in the hot blast heater, a pipe was installed in the side near the top of the blast furnace shaft. A valving arrangement was devised to prevent the gas from escaping through the top, while at the same time permitting the charging of the raw material into the furnace. In England, the most popular such device was called a *cup and cone*. In the United States it was a *bell and hopper*, since the closing valve resembled a church bell that closed against a large cylinder that temporarily held the raw materials and was therefore called a hopper (see figure 4.5). The gas was removed from the furnace in a rather large pipe, usually called the *downcomer*, from which it could be used for heating the blast air or at the boiler house for making steam. Blast furnace gas is a low heating-value fuel of about 90 to 100 BTUs per cubic foot that eventually replaced the coal used for this purpose. Because of the problems involved with heating air flowing directly through pipes, *hot blast stoves*, mimicking the regenerative brick-style heat exchanger of Siemens design, were later developed. Although there were many variations, some of the principal designs that evolved were by Whitwell and Cowper. These devices displaced the cumbersome, breakdown-prone heat exchangers that used cast iron pipes for heat transfer. A Cowper's stove (one of the popular basic designs that is still in use), for example, consisted basically of two parts: a combustion chamber and a heat sink, or what became known as the *checkers*. It looked like a very tall cylinder with a dome on top. (Eventually they were almost as tall as the blast furnace itself.) The effective use of a stove at a blast furnace was a two-stage operation, one that involved first heating the stove and then heating the blast air. With the stove *off blast* or isolated from the air going into the stove and thence the blast furnace, blast furnace gas extracted from the top of the furnace through the downcomer was then introduced into the stove through a burner located near its bottom, along with enough outside air to burn the gas. The gas burned on an upward pass through the combustion chamber of the stove, where it then made a 180° or "U-turn" in the open chamber at the top of the stove (the reason for the dome) and passed downward through the checkers, heating the firebrick from which they were assembled. The spent gas then passed into a channel and out to a chimney, which provided the draft. This style was frequently called a two-pass stove, one pass up and one down. The checker bricks were heated to nearly 2,000°F or more depending on the era (but only 1,040°F for the early Cowpers[17]). Upon reaching that temperature the gas and air for combustion were shut off and the valve at the connection to the chimney was closed, isolating it from the gas system. The cold blast air valve was then opened to introduce air into the stove where it would increase in temperature as it passed over the newly heated checkers. The hot (incandescent red) air then passed out through the now open hot blast

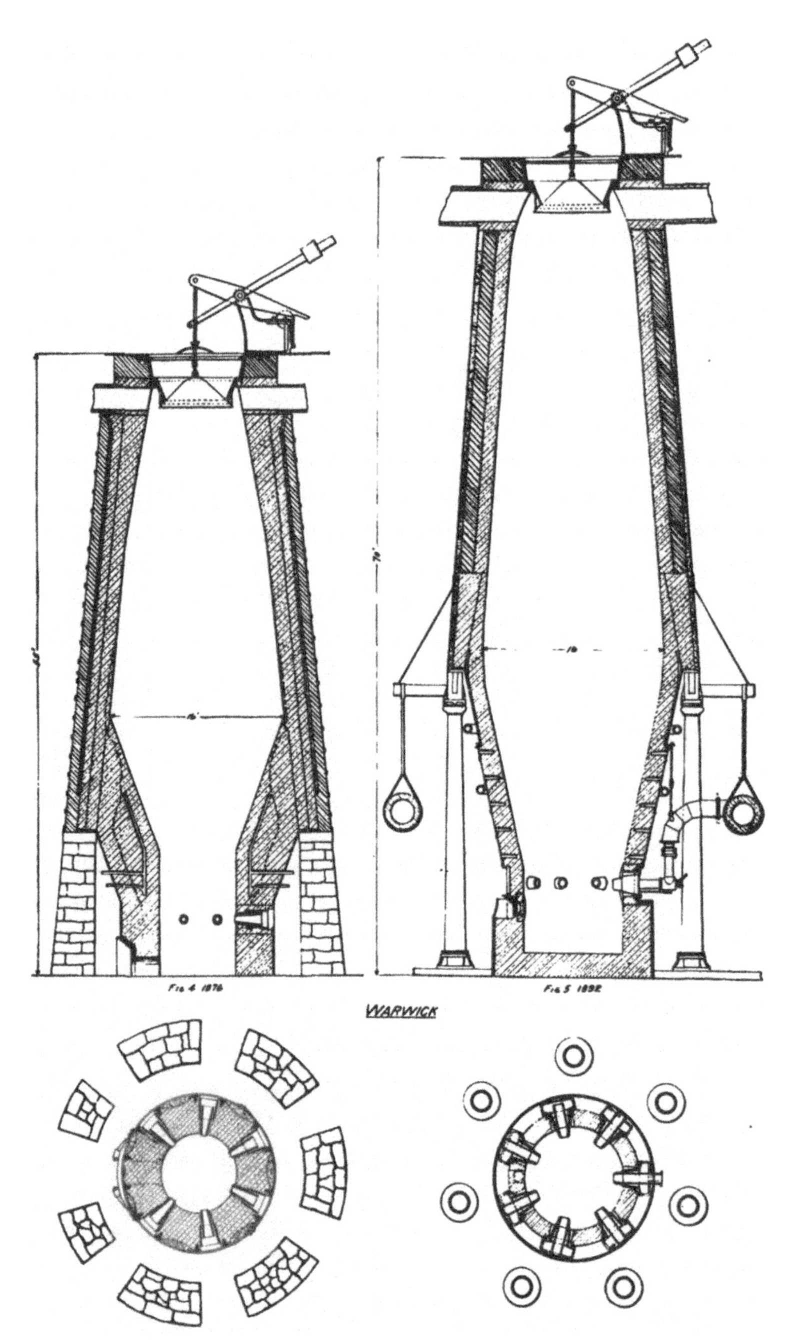

Figure 4.5. Cross-section of a blast furnace using the bell and hopper, also known as cup and cone, type gas lock. This permitted the collection of furnace gasses through the horizontal openings near the top of the furnace shaft located at the level of the bell. This also illustrates how materials could be removed to make the furnace lighter by replacing stone abutments with posts and adding cooling plates to the outside of the shaft and the addition of circulating pipe for the air blast called the bustle pipe.

Engineers and Engineering, Philadelphia, 1902.

valve and into the furnace. A plant needed at least two stoves of this type, one heating the air and the other heating the checkers. Three were even better, but most plants built four stoves per furnace: one on line; one heating; one in the process of changing, and one as an on-line backup or off line for cleaning or maintenance. With four stoves the cycle *on blast* was about an hour. If two of the four stoves were on blast, usually one was changed every half hour, but each was still on line for about an hour. Other operating regimes were practiced.[18] Some plants had three-pass stoves, easily identified by a stack mounted on top of the stove dome, with the path of the air up-down-up and out the stack or to the blast furnace.

One of the positive side effects of the hot blast was increased iron production. This meant more iron could be made in a furnace without increasing the size. Additional tuyeres were installed on the taller anthracite and coke furnaces, as many as twenty or more, but this was found to be excessive. The hot blast from the stoves was distributed to the furnace in a refractory-lined pipe, called the *bustle pipe*, that encircled the hearth. Because the air was red hot, the tuyeres were fitted with water-cooled pipes that eventually were cast with internal cooling ports to counter the detrimental effects the heated air had on the metal. Additionally, the ports through which the tuyeres were inserted into the hearth were fitted with castings called *coolers* that likewise had internal passages for cooling water to pass through. The danger of having all of this water in close proximity to molten iron collecting in the hearth is twofold: if there was an undetected water leak, it could chill the hearth, freeze the iron, and extinguish the furnace; and any time that molten metal covers water, it can cause an explosion. Steam explosions of this type sometimes were extremely violent.

The progress made in making iron, first by using charcoal, then coal, and finally coke was largely dependent on learning how to address the idiosyncrasies introduced by each fuel. The most obvious of these were the pressure and temperature requirements of the blast air. All three materials contained the prerequisite carbon to reduce the ore. The arrow of history, though, does not point directly from one to the next. It is a convoluted path frequently folding back on itself. As previously stated by Keir, "charcoal was the most desirable fuel because it made the purist iron," but it required vast tracts of land and large forces of labor. It was a fragile material, and blast pressures had to be kept low because of its fragility. Also, it could not support very well the heavy burden needed for large production, which required taller furnaces (for size comparison, see figure 4.6). Coal and/or coke were partially substituted for charcoal on occasion, but without much success. The use of coke, as developed by the Darbys, had challenges as well. It required higher blast pressure to produce iron, and the coal that they used had to be capable

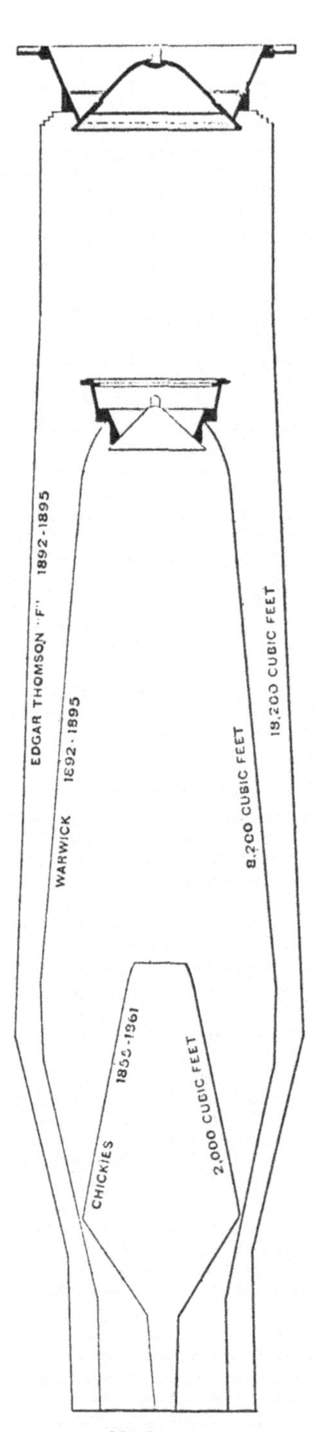

Figure 4.6. Illustration gives a comparison of the relative sizes of charcoal (1840 Chickies), anthracite (1845 Warwick) and coke (1892 Edgar Thomson "F") furnaces at approximately the same era.

From Iron Age, Vol. 57, 1896.

of coking. Sulfur, volatiles, and ash contained in the coal could contaminate the iron and had to be controlled. Sometimes, good ideas proved flawed and needed to be revisited after being given time for complementary technologies to catch up with them. For instance, bellows or inefficient wooden tub piston pumps driven by antiquated water wheels provided the blast for many furnaces. The invention of more reliable, higher-pressure, steam-driven blast engines with compound metal piston pumps was only a relatively short time away. The Darbys of 1740 were beset with the problems caused by a cold blast on their coke furnaces because hot blast technology was almost a century away. Starting in the 1830s, several facilities in Pennsylvania succeeded in producing iron using anthracite or hard coal alone as a fuel. Anthracite coal as mined was typically about 90 percent carbon. By using the hot blast and a higher blast pressure, the tendency of this type of coal to crumble and resist the flow of air was overcome. Large amounts of good-quality iron could be smelted without the need to produce charcoal. "With the beginning of the use of anthracite coal in her blast-furnaces, the distinctive charcoal era in her iron history [of Pennsylvania] may be said to have come to an end."[19] Several furnaces in Pennsylvania and in nearby Youngstown, Ohio, made iron directly from bituminous or soft coal, but it was proven impractical because of decreased strength and volatility as well as the amount of impurities introduced.

> The density of blast [pounds per square inch pressure] depends strictly on the quality of fuel . . . soft charcoal with ½ to ⅝ of a pound pressure . . . hard charcoal, with ⅝ to 1 pound pressure. The best wood charcoal will not bear more than this density. Raw bituminous coal, or coke is worked to advantage with 2½ pounds to 4 pounds pressure, and anthracite should have at least four pounds . . . As the effects of hot blast may be in some measure produced by higher densities, the best results must, as a matter of course, be obtained when pressure and temperature are so regulated as to work the ore with the smallest amount of fuel. We are not informed what density cold blast anthracite will bear; but we know strong coke will bear 6 pounds, hard charcoal 1 to 1½ pounds, and soft charcoal to ¾ and 1 pound.[20]

While much of the development and transformation of iron-smelting technology was happening elsewhere, Pittsburgh was establishing itself as the (wrought) iron and steel manufacturing center of America during the 1830 to 1850s era, without the benefit of having its own blast furnace. The iron industry of Pittsburgh, which became the most important iron-manufacturing center in the nation, did not exist in the eighteenth century, although this is not perfectly accurate. In 1792 George Anshutz built a small stone furnace on Two-Mile Run, now the Shadyside neighborhood of Pittsburgh, which was largely devoted to casting stoves and grates. "In 1794 the fire of the furnace lighted up the camp of the participants in the whiskey insurrection." The Anshutz or Shady Side Furnace was shut down and abandoned by the end of 1794 due to an insufficient supply of ore.[21]

During the charcoal era in the United States, the furnace went to the supply of ore, not vice versa. The closest supplies of iron in Western Pennsylvania near that time came from charcoal furnaces located around Uniontown and Connellsville in Fayette County about thirty to forty miles to the southeast of Pittsburgh, where the first furnaces in the region were concentrated. Not much of that iron production made its way to Pittsburgh, and with few exceptions these furnaces were largely gone by 1850. There were also coke furnaces on the Allegheny River at Brady's Bend, about fifty miles northeast of the city, but by 1850 these, too, were out of operation. In large part, Pittsburgh seemed to be reliant on blooms from the Juniata Valley. Starting in 1829 Pittsburgh was consuming six thousand tons of billets and 1,500 tons of pig, and its use generally grew from that point.[22]

> The high value of the Juniata iron [from central Pennsylvania] enabled it to stand the expense of the rough journey over the mountains to Pittsburgh. The customary method of transportation was to bend the bars of iron into a shape of an inverted U and put them over the backs of horses. These traveled in packs of fifteen, each horse carrying 200 pounds. The whole cavalcade was under the guidance of two men. Afterward, when the Pennsylvania Canal was completed, the Juniata iron went by water to Pittsburgh.[23]

In the winter months the iron was hauled on sleds to Johnstown, and with the spring thaw it floated down on the Conemaugh, Kiskiminetas, and Allegheny rivers.[24]

CLINTON, ISABELLA, AND LUCY

Having started operation in 1845, the Clinton Rolling Mill owned by Graff Bennett & Company[25] built a blast furnace on the south side of the Monongahela River in 1859 on the opposite shore of the river of what is now downtown Pittsburgh (see figure 4.7). It was Pittsburgh's first successful blast furnace. It went *on blast* at the end of October that year[26] and was described in 1863 as, "'simply a jacket' of boiler iron lined with fire brick. It was fifty feet high and twelve feet 'bosh.' [Actually 45 ft. by 12 ft.] The make of iron was twenty tons in twenty-four hours."[27] Being a first in a city that had already developed a reputation for trailblazing during the expansion of the early American iron and steel industry was a notable accomplishment. Clinton was designed to use Pittsburgh Seam coal mined from the hillside directly above

Figure 4.7. This is a "modern" photo of Clinton Furnace from c.1905. The furnace was equipped for skip charging by this time. Although it was a pioneer in Pittsburgh ironmaking and Connellsville coke usage, it eventually suffered from poor management and never had a great impact and was demolished by c.1927. Library of Congress collection.

the plant (Coal Hill) and coked on the premises. The mine portal and an inclined railway to the plant are clearly visible on a three-dimensional lithographic map of the city from 1875 by Krebs. The furnace was *blown in* on the last Monday of October and "was run for about three months, when, the coke made in this way not proving satisfactory, it was blown out." Not that anyone then realized, but what happened next began a sequence of events that forever changed the world. The company secured a supply of Connellsville coke that was made on the ground in pits at the Fayette Coke Works, located on the Baltimore & Ohio Railroad.[28] The furnace was blown in once again. It operated better, and "the result was so satisfactory that 30 ovens were built in 1860 and arrangements were made to secure a continued supply."[29]

It would be incorrect to assume that Connellsville coke was not used to make iron before this time. In 1835 the Franklin Institute offered a gold medal to anyone making not less than twenty tons of iron in a year using bituminous coal or coke. F. H. Oliphant claimed the medal in 1837. He made about one hundred tons of coke iron at his Fairchance Furnace near Uniontown, Pennsylvania. It remains doubtful that he ever received the award. He never made any more coke iron and turned again to charcoal. Also, Isaac Meason made wrought iron from ore in what is known as a *refinery* (furnace), sometimes called a *finery fire*, using Connellsville coke that he made on the ground.[30]

> These early efforts in the use of coke in blast furnaces [universally] were not very successful. Probably this came from the imperfect methods of making coke and the insufficient blast to the furnace. The latter was, perhaps, the most retarding cause in the early efforts of smelting pig iron with coke."[31] "It was not until the Baltimore and Ohio [Railroad] was completed to [Pittsburgh] and Connellsville coke had been used successfully in the Clinton furnace at [Pittsburgh], that its value as a furnace fuel was fully demonstrated and the foundation laid for the demand which has resulted in such an unprecedented development of coke manufacture in the Connellsville coke region.[32]

Statistics from 1874 and 1880 indicate that Clinton's production was twelve thousand tons per year (about thirty-three tons per day). The ore used was primarily from the Lake Superior region with the balance from Missouri.[33] It is possible that Lake Superior ore was also used in 1859, as the first recognized use of this ore was at Sharpsville Furnace, Mercer County,[34] about seventy-five miles north of Pittsburgh in 1853 to 1854. Its use at Clinton at the beginning of operations in 1859 and early 1860 is not certain.

Coke serves three functions in a blast furnace. It acts as a fuel, a source of energy used to break the bonds in the ore. It acts as a reductant, in this case a supply of carbon to combine with the oxygen of the ore to free or reduce the ore to iron. It acts as a platform or matrix to support all of the raw material pressing down from the tall shaft above, while at the same time providing sufficient passageways for the gasses from the hearth to pass up through the stack (*permeability*) and start the reducing reactions. The last is an extremely important characteristic that is often overlooked. Lower-strength coke, like charcoal, could not withstand the pressure from above and would crumble and obstruct the transmission of air/gas through the furnace. Also, too low blast pressure, as mentioned previously in reference to the Brady's Bend furnaces, had a similar effect, and the gas could not pass up through the stack. While it may seem strange, air is the single most-used raw material per ton to make iron, both in terms of volume and weight—more tons of air are used than iron ore, limestone, or coke per ton of iron produced. Using some early twentieth-century data, for example: "To produce a ton of iron under favorable conditions requires about 2 tons of ore, ½ ton of flux, 1 ton of fuel and about 4 tons of blast."[35] As it is likely that Clinton had enough blast pressure when using local coke, and, as the limestone and ore would likely have been the same, it would appear that the introduction of Connellsville coke was solely responsible for improved production, and its use was one of those "being in the right place at the right time" occurrences. "Connellsville coke was found to work admirably as a fuel for blast furnaces in connection with a powerful blast and high temperature."[36] Clinton's use of Connellsville coke forever proved the value of this product and, as you will see, its use advanced the further development of the American iron industry.

Ironmakers of this era cast their iron at the furnace into products like pots, grates, and stoves or into bars known as pigs. In fact, *pig iron* acquired its name from the method in which the bars of iron were cast. The iron was tapped or cast from the furnace through a long, shallow trough called the *runner* that was dug in the sand floor of the cast house. This had a number of equally spaced perpendicular branches or channels called *sows* about twenty to twenty-five feet long, off to one side. Each sow had a series of branches about three-and-one-half to four feet long and approximately four to six inches square called *pigs*. This arrangement of troughs or moulds used to shape the iron when it solidified looked to many like a row of mother pigs or sows feeding their litters of babies, so it was called *pig iron*. Another way to visualize it is as a series of combs one above another.

When operators cast the furnace, a *gate* or dam was located in the main runner just after each row or sow so that it would divert the molten iron into that particular sow or pigs. When that series of moulds was full, the dam would be removed and the iron flowed on to the next sow and so on, until the furnace was emptied or the pig bed was full. Some sand was thrown over the metal after each row was filled to facilitate forming a skin over the pigs to prevent a breakout of molten metal. Because of the intense heat, men had to walk around in the area wearing shoes with one or more inch-thick wooden soles attached to them. The wood was simply a method of creating a barrier or a form of insulation that would protect their feet from severe burns, if they had touched the molten or red-hot iron in the moulds. Because it retained heat, leather shoes could actually exacerbate the wound. Sometimes water was sprayed onto the pig bed to facilitate more rapid cooling after the skin was formed. Then the backbreaking, laborious part of the job began. The men had to go into the pig bed with crowbars and sledgehammers and lift the pigs with long bars, while another member of the crew broke them into approximately one-hundred-pound pieces with a sledgehammer. Next, the solidified iron had to be picked up by hand and loaded into cars or stacked somewhere out of the way.

> At the Steelton [PA] furnaces [Negroes] and Hungarians made up the pig-bed force. At the Lackawanna Iron and Steel Company furnaces at North Cornwall, American whites with a few Hungarians completed the force. . . . The work calls for great physical strength and endurance. The intellectual demands are nil. "It's lift pig and that's about all," . . . "Gorilla men are what we need," is the way one employer put it.[37]

The work was not yet done. The pig bed had to be cleaned and straightened out. More sand had to be added to the cast house, if needed, and empty moulds made for a new pig bed. The main runner and sows had to be redone with new gates added. Patterns were laid in place of the pigs, and sand would be shoveled around them and tamped down. The patterns would then be removed. This sequence was repeated until the requisite numbers of rows were finished and the new bed was made ready. Most factories had two or more pig beds; one of the beds was being cast, or broken and loaded, and the other was being made ready for the next cast. The beds had to be prepared and available for at least two, but as many as six, casts per day. Failure to be prepared could result in serious damage to the furnace. With nowhere to go, iron would begin to fill and damage or destroy the tuyeres and hot blast piping, breaking the water piping and resulting in the loss of the blast and leaving water leaking into the furnace. This was a filthy, hot, demanding, strenuous job for low pay.

> The number of pig-men differs with the casting capacity of the furnace—usually from 6 to 15 pig-men are employed . . . Wage rates range from 18 cents to 21 cents per hour [c. 1917] . . . In quite a number of instances the flat rate of 12 cents a ton is paid to the pig-bed force for every ton of iron cast. That is, if 20 pig-bed men cast 500 tons, they would receive $3.00 each.[38]

There was a bit of ambiguity in the meaning of the word *ton* in these times. In America a ton was two thousand pounds, or a short ton versus the long ton of 2,200 pounds (one thousand kilograms) in England. A ton of billets or scrap was 2,240 pounds. Stranger still, a ton of pig was 2,268 pounds in recognition of the fact that pig contained sand from casting.[39] At Clinton the process of pig casting was well documented through a series of stereographs produced in the very early twentieth century. Some of those photos can be seen in figures 4.8 through 4.10.

Figure 4.8. Making a pig bed, Clinton Furnace, Pittsburgh c. 1905. Wood block patterns are properly placed, sand was shoveled around them and tamped to compact it, and the patterns were removed. This formed the pigs. The sows and runner were made by hand. Library of Congress collection.

The labor force that operated a typical furnace of this time, excluding the pig-bed men, consisted primarily of two groups: *stockers* or raw material handlers, and the men connected with the physical control of the furnace. The stockers were themselves composed of two groups: *bottom fillers*, or men whose job it was to retrieve ore, coke, or limestone from a stock pile, hand fill and then weigh large, two-wheeled barrows with the raw materials that were then put on an elevator to be hoisted to the top; and *top fillers*, or men who took the loaded barrows the short distance from the elevator to the bell and dumped them there. At the appropriate time, they would operate the mechanism that lowered the bell to fill the furnace with the feedstock. For a furnace that produced from one hundred fifty to two hundred tons of iron per day, twelve to fourteen bottom fillers and four to six top fillers were required. Both occupations were strenuous and boring. To produce two hundred tons

Figure 4.9. Casting a furnace into a pig bed at Sloss Furnace, Birmingham, AL, c.1906. The molten iron is flowing into the bed on the left, while broken pigs can be seen littering the floor on the right. The gates or dams can be clearly seen in the runner from the furnace, center.
Library of Congress collection.

Figure 4.10. Pig bed men toting broken pigs and loading them into rail cars at Clinton Furnace, Pittsburgh, c.1905. Given the size of each of the bars, every man is carrying upwards of 200 pounds.
Library of Congress collection.

of iron at this time required retrieving and loading approximately four hundred to six hundred tons of ore, three hundred to four hundred tons of coke, and about one hundred tons of limestone, or about 800 to 1,100 total tons of raw material every day. All were loaded and moved by hand. Fire from blast furnace gas igniting was forever a danger to the men on top, but not nearly as bad as it could be when it didn't ignite. Furnace gas was composed of a high percentage of carbon monoxide gas, which is colorless and odorless as well as toxic. Carbon monoxide is found in our home heating units, too, but not at the levels found in the furnace. The bell was never a perfect seal, and some gas was always leaking out. On occasion men were overcome by the effects of the gas while standing in the open air, especially when the bell was lowered and there was a sudden onrush of gas from the furnace. Sometimes as they passed out, they would fall through the open bell into the furnace shaft. Another common occurrence, called a *hang and a slip*, was when the coke would bridge the furnace shaft (a scaffold), greatly reducing or stopping the flow of gas through the furnace. Most often the hang would break quickly with no consequences. Yet on occasion a cavity would form below the *hang* as the free material continued to descend and the pressure would build there. When it broke, the material that had been suspended above would break free and rapidly descend into the shaft or *slip*. The sudden release of pressure was like a balloon bursting, and sometimes incandescent material was ejected from the top of the furnace, burning those who were working there. Though rare, there was sometimes such a forceful explosion accompanying the slip that part (or even the entire top) of the furnace was blown off, killing and maiming those present.

The group that had actual control of the furnace proper was comprised of occupations with titles including the *blower,* the *keeper* and his *helpers*, the *monkey-boss* and *cinder-snappers*. The blower was the boss. He had to understand the intricacies of furnace operation and know how to identify and understand how to resolve problems, preferably before or even as they arose. The keeper and helpers were responsible for maintaining the tap hole and the iron runners. They would open and close the tap hole when the iron was cast from the furnace, and also prepare and maintain the runners to receive the flow of iron. Before mechanical drills were invented, men had the strenuous job of opening the tap hole. They would stand at the hole with bars (hand drills) and use sledgehammers to open it by hand:

> Now six men with a long steel bar are starting to break through the two or three feet of clay with which the "tap hole" of the monster furnace is plugged . . . For ten minutes with strong sledge blows the tappers struggle to break through the plug of burned clay . . . And now a shout and the strong red glow throughout the cast house tell us that the tap hole is open and the iron is running down the main channel of the sand bed.[40]

After the use of mechanical drills became common, the drill was frequently used to open the hole most of the way, then the men went in with hand drills and sledgehammers to finish opening the tap hole so that the drill wasn't damaged by molten metal. Men had the sense to get out of the way—most of the time. Worse still was closing the tap hole. After the iron was cast, a shower of sparks from hot coke, slag, and molten iron continued to spew from the tap hole. The keeper would have one of his men attack the hole while dodging the ejecta from the fireworks display. At the same time he used a bar that had a ball of mud attached to its end to insert a plug into the hole. This often required a number of attempts, and after the first success, the hole had to be filled the rest of the way with enough mud to hold back the iron until the next cast. Men had to do all of this while working next to a runner that was still glowing with an incandescent red heat from the just-completed cast. The monkey-boss and cinder-snappers dealt with slag. Slag was primarily composed of molten limestone that was used to carry away undesirables like sulfur and ash, which would be detrimental to the iron. Slag was removed from the furnace through a hole that was a few feet higher in elevation than the tap hole and off to the side. Because the slag was lighter, it floated on top of the iron. The hole to remove the molten slag was referred to variously as the *slag-notch* or *cinder-notch*, or the *monkey* depending on the plant.

The above-mentioned crew was only a portion of the number of men required to successfully operate a plant. There were also numerous laborers needed to unload raw material from cars and to clean the enormous amounts of debris that was generated. Engineers were employed to operate the boilers and blow engines, and there were stove tenders, millwrights, pipe fitters, riggers, oilers, bosses, and so on.[41]

Starting with Clinton's success in 1860, more stacks were planned and built in Pittsburgh. The demands imposed by the Civil War certainly gave it additional impetus. In 1861 Jones and Laughlin put two stacks (45' × 12') on line called the Eliza Furnaces. They were located almost within sight of Clinton, about two miles upstream on the opposite shore of the river. In 1862 to 1863 two more were built, later called Edith, for the Superior Furnace Company (45' × 12'), down the Ohio River at Manchester. The years 1864 to 1865 found two more constructed for Shoenberger (62' × 13½') near the Pennsylvania Railroad Depot on the Allegheny River. A single Soho Furnace (65' × 18⅔'), located between Clinton and Eliza, was put on line in 1872.[42] The trend was to increase furnace size, indicating a desire for increased output.

While in early 1859 Pittsburgh had no blast furnaces, the end of 1865 and the conclusion of the war saw seven, all within walking distance of each other. Soho was not the only plant to go into production in 1872; three additional stacks were constructed, two for the Isabella Furnace Company

(78' × 18' and 75' × 20') in Etna and one for the Lucy Furnace Company (75' × 20') at 51st Street in the Lawrenceville section of the city.[43] These two new plants were across the Allegheny River in sight of one another. The three new furnaces, some one-and-one-half times the size of the original group of 45' × 12' Pittsburgh stacks, were modeled after the larger, higher-production English furnaces of the time. A consortium that included the Spang and Chalfant Company, Painter and Sons, Henry Oliver, and Graff Bennett & Company, who were the proprietors of Clinton Furnace, owned Isabella. Lucy was established by four men who were already engaged in wrought iron manufacturing at their Upper and Lower Union Mills, where they would use the production from Lucy in their puddling furnaces and rolling mills. This group was made of Andrew Kloman, the owner and operator of a forge on Girty's Run in Millvale before owning the Union Mill; Henry Phipps, a partner of Kloman at Girty's Run and in the Lower Union Mill; and finally a pair of brothers, Tom and Andy Carnegie. This was their first venture into the iron smelting business. The partners in Lucy Furnace had been invited to join the Isabella group, but on the advice of William Coleman, Tom Carnegie's father-in-law and a man whose judgment and business acumen they trusted, they declined to participate. Coleman related to the group that "it would be better for them to build one furnace themselves than to own one-seventh of two furnaces which would not be under their control or management."[44] Isabella (see figure 4.11) was named for Isabella Herron, the sister of a member of the firm at Spang and Chalfant, while Lucy (see figure 4.12) was named for the daughter of Coleman and the wife of Tom Carnegie.[45] Andrew Kloman was, according to Andrew Carnegie, a mechanical genius. Henry Phipps was partnered with Kloman, but he was also a childhood friend of Carnegie, although he was closer in age to Andy's younger brother, Tom. Friends knew him as Harry.[46] Of the four original associates in the venture, Kloman was the first to leave following a bitter dispute that began as an issue with the ore supplies for Lucy. He sought revenge against Carnegie that ended only with his death in December of 1880.[47] Tom was known as a good manager—"Tom Carnegie's word is better than most men's bond"[48]—but under constant pressure he became an alcoholic and died from pneumonia in 1886, leaving only Phipps and Andy to reap the full reward that came with the formation of U.S. Steel in 1901.

Right from the start Lucy and Isabella produced an average of about fifty tons of iron per day, which was a pretty solid figure, even for established furnaces of that time.[49] Lucy was described in an 1873 article in *The Iron Age* as

> built by Messrs. E. J. Baird and Wm. Tate, and went into blast on May 2, 1872. It is 75 feet high by 20 feet diameter of bosh. Like most Western furnaces, it is an iron cylinder lined with fire brick, with an independent iron gas flue, around

Figure 4.11. Isabella Furnace, Etna, PA. These two stacks were owned by a consortium that included the Graff Bennett and Spang and Chalfant companies. The start of these furnaces, together with Lucy Furnace, which was located across the river in Pittsburgh, marks the beginning of a new era in the history of ironmaking, c.1873.

Courtesy of the Hagley Museum and Library.

Figure 4.12. The Lucy Furnace, 1873, only one year after the start of operations. The photo is definitely prior to 1877 as there is only a single stack. This was the first furnace in the world to produce 100 tons of iron in a single day, an unimaginable feat for the time. US Steel News, November 1937, p. 25.

Courtesy of the Hagley Museum and Library.

which winds an iron stairway by means of which access is had to the top of the furnace. The fuel and ores are carried to the tunnel head in barrows by means of a pneumatic lift, from which they are run under cover of an iron roof to the top of the stack and dumped by hand . . . The machinery of the works is of the best quality, though of a very different character from that usually seen in the East . . . There are three excellent blowing engines by Messrs. Mackintosh, Hemphill & Co., Pittsburgh . . . The capacity of the furnace is about 550 tons per week, taking the average of the seasons. The ores used are mostly Lake Superior . . . The fuel is coke made from the slack of the bituminous mines near Pittsburgh, coked at ovens located at Carpenters Station [later renamed Ardara] on the Pennsylvania Railroad about 19 miles distant . . . the consumption in the stack is only about one and a half tons [of coke] to the ton of pig iron made.[50]

And, from other texts:

The Isabella furnaces were built by Benjamin Crowther, and each has independent air hoists, stoves, blast mains, &c. In the engine house . . . stand 5 short-stroke blowing engines, with balanced slide valve, by Mackintosh, Hemphill & Co. . . . The stroke is 4 feet, and the diameter of the blowing cylinders is 84 inches and the steam cylinders 34 inches . . . Immediately in rear of the furnaces stand 10 hot blast stoves, with pipes 4 by 16 inches in the two diameters, corrugated inside, and 14 feet long. The blast is heated to 1000°, and each furnace is blown through 7 tuyeres, 6 ½ inches in diameter, with a blast of not less than 7 pounds to the square inch. Adjoining the stoves to the rear is the stock-house, 240 feet long, 75 feet wide, and 34 feet high, in which 10,000 tons of Lake Superior ore can be stored, and coke, from the company's own works at Blairsville, is dumped into hopper bins, from which it falls into barrows."[51] "Hoisting of materials is accomplished by means of two pneumatic lifts, one for each furnace. They are located in the rear of the furnaces between the two groups of ovens [hot blast stoves] . . . their base being on the general level of the stock house . . . Each [hoist] consists of a simple cylinder of cast iron, 92 feet long, and 36 inches in diameter . . . in the interior of which is a loosely fitting piston with balance weight . . . The hoist is worked by admitting air from the blast main into the cylinder . . . by means of valves, without the use of air pumps . . .

At each ascent, a barrow containing 500 lbs. of coke, and one containing 900 lbs. of ore or limestone are taken up. 120 trips are made per day with each lift. The actual time required for putting the loaded barrows upon the cage, raising them to the top, emptying them, and returning them to the bottom, is about 1 minute and 40 seconds.[52]

Coke was obtained from: Two miles east of the town [Blairsville, PA about 60 railroad miles distant], on the [Western Pennsylvania] railroad, are the Isabella Coke Works, consisting of two hundred ovens, and employing as many men. These works extend along the Conemaugh [River], and at night present a strikingly grand appearance.[53]

One of the more startling realizations about either furnace's operation as presented above is that neither one used Connellsville coke early on, even though Ardara and Blairsville were relatively nearby that important district. This assertion was especially true for Isabella, whose part owner, Graff Bennett, pioneered the use of the Connellsville product at Clinton Furnace. A later *Iron Age* article notes that the coke requirement for Lucy in excess of the amount made at Carpenters was gotten from Connellsville.[54] The fact that Carnegie's facility used slack coal (normally considered a waste) from Pittsburgh mines to manufacture coke highlights a practice that became an enduring theme of his company's operation, that of closely watching the costs, and coincidentally, efficiently using materials whatever they may be. Both facilities were similar and had comparable equipment sometimes made by the same manufacturer, so a friendly rivalry soon developed between the two. Very shortly though, this rivalry ripened into a battle for dominance of the world production record for blast furnace output. Initially, the type of iron output was probably similar too, because they were both feeding puddling furnaces and rolling mills, but by 1875 Lucy was producing Bessemer iron to be converted into steel, while Isabella's focus was still on forge iron and foundry pig.

Before the Edgar Thomson plant was built, Bessemer-grade pig iron that was smelted at Lucy was sold to the Pennsylvania Steel Company in Harrisburg. According to Bridge, because there was a penalty for each one-hundredth percent of phosphorus in the iron over a set limit, Phipps hired a chemist to keep track of Lucy's production.[55] However, by Carnegie's recollection, they hired Dr. Fricke due to more general problems with the performance of Lucy. Operations were very erratic. Their manger controlled the furnace by "rule of thumb" or "seat of the pants" methods in addition to physically bullying the employees. Carnegie also suspected that he used "divination" or "instinct." They replaced him with an inexperienced but trusted young shipping clerk, Henry Curry, and found a chemist to assist and guide him, while Phipps watched the overall operation. Because the Lucy Furnace Company was a relative latecomer to the iron smelting game, much of the readily available high-quality raw materials were already spoken for and their reserves committed to others. Therefore, Lucy was saddled with having to use a wide assortment of coke, limestone, and ore supplies. Yet by having a chemist in house they found that:

> Iron stone [iron ore] from mines that had a high reputation was found to contain ten, fifteen, and even twenty per cent less iron than it had been credited with. Mines that hitherto had a poor reputation we found to be now yielding superior ore. The good was bad and the bad was good, and everything was topsy-turvy.

> Nine tenths of all the uncertainties of pig-iron making were dispelled under the burning sun of chemical knowledge. At a most critical period when it was necessary for the credit of the firm that the blast furnace should make its best product, it had been stopped because an exceedingly rich and pure ore had [unknowingly] been substituted for an inferior ore—an ore which did not yield more than two thirds of the quantity of the other. The furnace had met with disaster because [based on the rule of thumb for the assumed quality of the ore] too much lime had been used to flux this especially pure ironstone. The very [unexpected] superiority of the materials had involved us in serious losses.[56]

While other companies pondered how Carnegie could afford the extravagance of a "professor," Carnegie wondered how they could not.

When making lower grades of iron, some companies would often use puddle furnace cinder as a partial replacement for ore as a feedstock because it was cheaper. Cinder was the slag that was expelled from the production of wrought iron. This slag contained a considerable percentage of iron. The problem was that it also had an appreciable amount of phosphorus as well. If proportioned correctly with a good grade of ore and the purchase price for the cinder was reasonable, and if the customer would accept higher phosphorus in this iron, then using cinder could be a profitable undertaking. By using this waste generated at the Union Mills, Lucy could in essence get that portion for free, as it came from within the company. It was common practice in the local area to use cinder to make a product called "mill iron." But Phipps realized that there was yet another potential source of iron at the mills: the fine dust (flue cinder) that was carried from heating furnaces into the stack and collected there. At most plants it was a nuisance because it had to be cleaned from the chimney where it sometimes blocked the passage of air. Locally, it was usually dumped in the river as waste. Phipps

> had some of this cinder analyzed [by their chemist], and found it as rich in iron as the puddle-cinder . . . and carried less than one-fifth the amount of phosphorus . . . the firm sold its puddle-cinder [from the Union Mills] through brokers at $1 and $1.50 per ton . . . and in the same way bought [for Lucy] . . . flue cinder for fifty cents a ton.[57]

In 1873 Thomas Whitwell, a patent holder for a popular design of hot blast stoves, proposed to the owners of Lucy that they modify the shape of the bell to change the distribution of raw materials in the furnace. He claimed that this would reduce wear on the lining and segregate the contents to produce a better flow. Whitwell convinced the partners at Lucy to adopt his idea by creating a scale model of Lucy that had a glass front and was set up in the furnace yard. By looking through the glass of the model, the effects that the modified bell had on the charging operation could be seen. Phipps and Andrew Carnegie were among those present for the experiment. Bridge reports that the modifications were made to the furnace that resulted in improved output, but another reference in about the same time frame indicated a return to a smaller bell. Whether the two are connected is not certain, but the point is that at Lucy, scientific methods sometimes were trusted during development of the ironmaking process.[58]

Soon both the Lucy and Isabella furnaces would gain recognition, first on a national and then on the world stage, as they alternately decimated world iron production records. Then not being satisfied by that, they began to attack each other. According to Wall,

> Average production for a furnace was considered to be . . . 350 tons a week. Both furnaces quickly passed that mark and by the end of the year [1872], the Lucy was averaging 500 tons a week . . . and the Isabella was close behind. . . . The next year both furnaces were breaking all records for both America and England. In 1873 the Lucy was producing an average of 593 tons a week, only to be passed by the Isabella with 612 tons. In October of 1874, Lucy jumped ahead with 642 tons, and late in the month there was great celebrating at 51st Street when the Lucy broke the record by producing over 100 tons in a single day. Then the Isabella on Christmas Eve passed that mark with 112 tons for the day.[59]
>
> . . . in 1874, when, for the first time in the history of ironmaking, the Lucy turned out a hundred tons of iron in one day. In England the news was received in silent incredulity.[60]
>
> As surpassing "anything ever before done in the world" the American Manufacturer of the 25th November, gives the return of the Isabella furnace No. 2, for the week ending November 20th, 1875, which shows a total make of 770 tons 660 lbs [or, more than a hundred tons per day average.][61]

This level was surpassed when Lucy had a weekly make of over eight hundred tons in 1878 and then in 1880 made over nine hundred tons. Not to be outdone, in February 1881 Isabella produced one thousand tons for one week, which was an astounding figure for that time, about three times more than what she made when she first started a little more than eight years before. Although these furnaces were mere babes, the media pronounced, "What will these Titans do next?"[62] "Trade journals carried the weekly totals as other papers carried baseball scores."[63]

A second furnace was erected at Lucy in 1877. The Edgar Thomson Works in Braddock began steelmaking operations in 1875 and required considerable amounts of pig for their operation in addition to the iron needed to keep the Union Mills running. Lucy No. 2 was built along the same lines as the No. 1 furnace. After all, it was a world record holder at times. The ancillary equipment was the same as well. There were four additional hot blast stoves with cast iron pipes, two additional blowing engines, and the same style of pneumatic elevator to hoist raw materials. Cast plates were added to the stack, which could then be sprayed with cooling water if necessary to address overheating in that area. It also had a gas washer added to the downcomer so that dust from the blast furnace gas could be sprayed with water to be collected and washed away.[64] Blast furnace gas was notoriously dirty due to the high velocity of the gas flowing through the furnace. It carried fine particles of raw material out of the stack along with it.

In order to describe the operation of these two furnaces and the others that would soon follow their lead, a new term became part of the industry jargon: *hard driving* or *rapid driving*. It was not necessarily a compliment, especially in England, where they worked to protect the furnace lining.[65] Hard-driven furnaces were worked for high outputs, which came at the expense of a decreased life of the refractory lining and could also have an additional cost in fuel. This revolution in ironmaking came about in part because operators at Lucy and Isabella recognized and then put into use the fact that the high production of the Struthers (Ohio) Furnace was attributed in part to the manager (Thomas Kennedy) having established a practice of counting the revolutions of the blow engine, instead of regulating the blast pressure.[66] Counting revolutions regulated the *flow* of air into the furnace instead of the pressure at the tuyeres. If a furnace was beginning to plug and/or resist the flow of air, the pressure at the tuyeres would rise. By following previous operating paradigms for regulating the pressure, logic would have directed them to compensate by slowing down the blowing engine (essentially reducing the number of revolutions of the flywheel) in order to reduce the pressure. For various reasons this was often the exact opposite of what was needed. Likewise, if the pressure were to drop due to channeling or some other reason, trying to increase the pressure by speeding up the blowing engine could exacerbate the problem. As long as the rate of flow and the mixture of raw materials being added to the furnace remained relatively constant, the amount of air required should have remained the same. Counting the revolutions of the engine guaranteed a constant flow of air. During each revolution, the air piston stroked up and down once. For both Isabella and Lucy, this meant a flow of about 154 cubic feet of air per stroke or 308 per revolution (as each company's blowing engines had the identical air piston diameter, seven feet, and stroke, four feet). Maintaining a flow of fifteen thousand cubic feet per minute (cfm) was simply a matter of counting forty-nine revolutions in the same amount of time. By using this method, a change of pressure at the tuyeres could now be used as a harbinger of developing problems. It also tended to reduce fluctuations in the operation. As is not uncommon, simple ideas that frequently are contrary to popular wisdom sometimes reap great rewards.

> The hard driving to which Americans resort . . . means a short life to the furnace. . . . the lining of an American furnace . . . lasts four years . . . in one case the lining of a British furnace lasted eighteen years. The Americans, however do not estimate so much on the time a furnace lining will last, as on the iron it will produce. "We hold that a lining is good for so much pig, and the sooner it makes it the better." . . . In spite of this greater cost of lining per ton of iron produced of the American practice . . . ironmasters of the United States believe in hard driving.[67]

"Less than 50 years ago the American blast furnace which would make 4 tons of pig iron in a day, or 28 tons in a week, was doing good work."[68] The world of this time was impressed and amazed with the wizardry of Isabella and Lucy, but their outpouring of iron only set the stage for what was to come at Edgar Thomson in the not-to-distant future.

Table 4.1. Pig Iron Production in the United Kingdom versus the United States, 1854 to 1872

*Year**	*United Kingdom (tons)*	*United States (tons)*	*Introduction of Furnaces at Pittsburgh*
1854	3,069,838	736,218	
1855	3,218,151	784,178	
1856	3,586,377	883,137	
1857	3,659,477	798,157	
1858	3,456,064	705,074	
1859	3,712,904	840,627	Clinton
1860	3,826,752	919,770	
1861	3,712,390	731,554	Eliza (2)
1862	3,943,469	787,662	
1863	4,510,040	947,604	Superior (2)
1864	4,767,901	1,135,996	
1865	4,819,254	931,582	Shoenberger (2)
1866	4,523,897	1,350,343	
1867	4,761,023	1,461,626	
1868	4,970,206	1,603,000	
1869	5,445,757	1,916,641	
1870	5,963,515	1,865,000	
1871	5,627,179	1,912,608	
1872	6,741,929	2,854,558	Soho/Lucy/Isabella (4)
1873	6,850,000	2,868,278	

*Table tonnage data compiled from: *The American Exchange and Review* XXVI, Philadelphia, 1875, 199.

Part III

THE MOVE TO MASS-PRODUCED STEELMAKING

Chapter Five

The Bessemers Arrive at Braddock

How a Bessemer looks to a veteran steelworker's eye I don't know. To a layman it looks like the egg . . . with one end cut off and gaping. It is a container of brick and riveted steel, twice as tall as the tallest man and supported near its middle on axles. It is set high up on a groundwork of brick. Into this caldron goes molten ore, fifty thousand pounds at a time. Through the iron is blown cold air—oxygen forced through the hot metal with the power of a giant's breath. Out of the egg in good time comes steel. It is a little short of pure magic.

One of the converters was in blow . . . the mouth of it belched flame . . . It was a terrifying sight, and hypnotic. I didn't want to look elsewhere, to turn my eyes from the leaping flame which towered thirty, perhaps forty, feet above the converter.

The roar was literally deafening; and little wonder, for here was a cyclone attacking a furnace in a brief but titanic struggle, a meeting in battle of carbon and oxygen cleverly arranged . . . Both carbon and oxygen would lose, each consuming the other, and men would be the winners by twenty-five tons of new steel.

The roaring continued. The red fire changed to violet, indescribably beautiful, then to orange, to yellow and finally to white, when it soon faded. "Drop," the boys call it. I saw the great vessel rock uneasily on its rack, moved with unseen levers by an unseen workman . . . The hellish brew was done.

Slowly the converter tilted over, and from its maw came a flow of seething liquid metal—Bessemer steel . . . pouring into the waiting ladle . . . Steel was being born in a light so blinding that one must wear dark glasses to look on it long. . . .

In perhaps five minutes the ladle was filled with the running fire. The bell on the locomotive rang. The ladle was pulled away, out into the darkness of the yard, and a sudden deep gloom settled down in the Bessemer shed. The devil's pouring was over.

It is the most gorgeous, the most startling show any industry can muster . . . enough to awe a mortal. No camera has ever caught a Bessemer's full grim majesty, and no poet has yet sung its splendor.[1]

Based on this abbreviated rendition of Bessemer operations at the Aliquippa Works of Jones and Laughlin Steel, taken from his 1939 work *Iron Brew*, author Stewart Holbrook might just have been that poet. Note: Twenty-five tons is representative of the capacity of larger converters (but not the largest) later in the Bessemer era.

BESSEMER'S NEW INVENTION

Henry Bessemer was an English inventor and experimenter of French heritage.[2] Bessemer's advancement in steelmaking came about through his effort to create the same rotation in a shell fired from a smooth bore cannon as that obtained from a rifled cannon. Rifling made gun and cannon fire far more accurate and longer ranged than a smooth bore due to the rotation it imparted to the shell. Yet rifled guns were much more difficult to manufacture. The English rejected his idea, but Bessemer successfully made an experimental version for the French government. Commandant Minié, the inventor of the bullet frequently called a "minnie ball" that was used to devastate so many soldiers' lives on both sides during the American Civil War, was present at the demonstration of the device. He congratulated Bessemer on his success but was not so much impressed as "he entirely mistrusted their present guns, and he did not consider it safe in practice to fire a 30-lb. shot from a 12-pounder cast-iron gun. The real question . . . was; Could any guns be made to stand such heavy projectiles?" This new challenge gave Bessemer the impetus to "improve" cast iron, ultimately resulting in a new method for making steel.[3]

Bessemer's experiments began by using a modified reverberatory furnace very similar to, but not identical, to one used for puddling. During his study, he noticed how the air that was drafted over and then later forced through the furnace had a decarburizing effect when melting pigs of iron. He then substituted blowing air directly into the liquid metal through a single ceramic blowpipe that was suspended above the vessel. In this experiment, in place of the reverberatory furnace the iron was kept in the molten state in a heated crucible. He recognized that there was a great deal of heat generated by this process and made a goal of not using fuel to maintain the metal in the liquid state. The crucible method that he used took about thirty minutes, and the iron would not have stayed hot enough for that length of time without additional heating. This eventually evolved into using a cylindrical vessel that had six tuyeres spaced around the perimeter of the bottom that became submerged by the molten metal after it was brought from another furnace, where solid pigs of iron were melted for the charge. Air with about ten to fifteen pounds of pressure was blown upward through the liquid metal. When the reaction was complete, the steel was tapped off to one side into a ladle. About seven hundred pounds of steel was made each time. Bessemer recognized the problems that were inherent in using a stationary vessel, so he developed a refractory-lined, ellipsoidal-shaped container open at one end that rotated (tilted up and down) on axles. This shape would be readily identified today as a Bessemer converter for making his self-described "cheap steel."[4]

Bessemer demonstrated his invention to George Rennie, then the president of the British Association for the Advancement of Science. Rennie recognized its importance and encouraged Bessemer to make a presentation to the association at a meeting in Cheltenham, scheduled for a week later. He agreed and prepared a paper titled "The Manufacture of [Malleable] Iron without Fuel." While Bessemer was eating breakfast with a friend prior to making the presentation, an ironmaker known to the friend but not to Bessemer approached and asked the friend if he knew "that there is actually a fellow come down from London to read a paper on the manufacture of malleable iron without fuel? Ha, Ha, Ha!"[5] Again there is this recurrent theme of a simple, successful idea of great significance that was met with indignation by "knowledgeable people" because it was contrary to popular wisdom. The Bessemer process heralded entry into the modern age of man, the age of mass-produced steel. Bessemer patented his process in 1856.

Bessemer was not a metallurgist. He was lucky in that he had accidentally used a good grade of Swedish iron ore feedstock with which to conduct his experiments. It contained low amounts of phosphorus and was high in manganese content.[6] By chance he had stopped the reaction at the correct time. The steel produced had very little carbon and was too soft. It was similar to wrought iron, but without slag or phosphorus problems. If held too long in the vessel it became overoxidized. If a Bessemer was overblown, it could make a mass of weak, useless, crumbly oxidized iron because it would in turn burn out the silicon, carbon, and other elements and finally burn the iron. Swedish ore was sometimes difficult and expensive to obtain, so when he switched to using iron prepared from local English ores that were low in manganese and high in sulfur and phosphorus (sulfur from contact with coal or coke during remelting), he did not understand what was happening and made steel that was referred to as *red short*, even if he managed to get the correct carbon content. Red shortness, sometimes called *hot short* or "rotten," describes a defect where the steel cracked, broke, or even crumbled when being worked or rolled at a red-hot temperature, certainly not a desirable characteristic when making steel. Also, when he used the lower-quality ores the steel suffered from "cold shortness," which was a defect where the steel cracked or broke when being worked cold. Phosphorus was the cause of cold shortness and sulfur contamination was responsible for red shortness,[7] neither of which could be eliminated in Bessemer's converter because it was made of an acid refractory (usually ganister in England). When he did use low phosphorus Swedish ore, the manganese it contained chemically fixed the sulfur, and it was flushed into the slag, eliminating that as a problem. It also helped to deoxidize the metal.[8]

According to Percy, "In attempting to produce steel by the methods specified by Bessemer, it has hitherto been found very difficult, if not impracticable, at least in this country [England], to ascertain with certainty when decarburization has proceeded to the right extent, and therefore the blast should be stopped."[9] David Mushet devised one of the methods for recarburizing cast steel for the crucible steelmaking process around 1800. His son Robert was an avid experimenter with iron and steel. Robert recognized that Bessemer's failure was due to two things: the iron needed to be recarburized, and there were some impurities present that needed to be eliminated. He also knew that these failings could be easily remedied through the addition of a measured amount of a material called spiegeleisen,[10] literally translated "looking-glass iron" due to its luster. It was also called spiegel-iron or just plain spiegel. Spiegel was composed of iron, manganese, and carbon. Later ferromanganese was developed with similar properties in different percentages. A product manufactured in America called Franklinite pig, together with spiegel or ferromanganese, were frequently referred to as the *triple compound* when used to adjust the chemistry of Bessemer steel because they all contained carbon, iron, and manganese. "Accordingly the plan now adopted is to decarburize perfectly or nearly so [remove all of the carbon], and then add a given proportion of carbon in the state in which it exists in molten spiegeleisen, the precise composition of which should of course be known."[11] The carbon was used to increase the carbon level (recarburize), and the manganese could bind sulfur and remove oxygen from the blown metal (deoxidize). Oxygen was present from bubbling air through the metal.[12] Mushet patented his idea in 1856, but he let it lapse through either his own or his representative's incompetence by failing to pay the required fees. Either way, the concept entered the public domain in England and was free for Bessemer to use without any encumbrances. "Mr. Mushet allowed this patent to become void in the third year. So far, spiegeleisen has given a value to Bessemer's process of making steel, which certainly it did not previously possess. . . . Mr. Mushet could hardly have been aware of the pecuniary value of the patent, or he would not have so gratuitously surrendered his rights."[13] Surely this method of making steel should be known as Bessemer-Mushet, because Bessemer's concept was impractical without it. Mushet practically (and Mushet's daughter actually did) begged Bessemer for money for the use of the concept, but while Bessemer received a knighthood and in excess of a million pounds over his life for the invention, he never officially recognized the critical value of Mushet's solution to his problem. Reluctantly, he did grant him a few thousand pounds through a stipend of only three hundred pounds per year.[14]

Bessemer applied for an American patent in 1856, but in 1857 William Kelly, an ironmaker from Eddyville, Kentucky, challenged his petition. Ironically, Kelly was a native of Pittsburgh. He claimed priority, and the testimony of witnesses provided sufficient evidence to prove that his "air boiling process" was used to make iron into steel as early as 1847, well in advance of Bessemer's development. Information provided in that testimony was several years after the fact, and a few of the details are subject to debate. Probably the most damming fact was that Bessemer did not contest the challenge within the allotted time, although it should be understood that at that time he was having a good deal of difficulty in England trying to identify and address the problems caused by phosphorus, sulfur, and oxygen contamination (prior to using Mushet's solution). So the patent was awarded to Kelly with the exception of that portion related to the machinery devised by Bessemer, for which he retained the patent. "The American experiments of Kelly, which antedated those of Bessemer, were successful only to a limited extent. . . . He blew air downwards into the metal, at low pressure, and in few and large streams; hence his iron chilled before decarburization was complete."[15] According to his biographer, after receiving the patent, Kelly was given an opportunity to perform a test at the Cambria Iron Works in Johnstown, Pennsylvania, with a converter he had built using iron that was smelted for him. In this test vessel, the air was blown up through the molten metal rather than down onto the bath as proposed by Kelly. Also, the equipment itself clearly resembled Bessemer's and was not the crude refractory cylinder that Kelly experimented with at Eddyville. "The blast . . . was so strong that the greater part of the contents of the converter were sent flying into the air in a brilliant tornado of sparks. This greatly amused the spectators, hundreds of whom were present."[16] But we must consider that this information as presented is a third-hand account of what occurred. While Bessemer was the initial source of laughter, immediately before he made the presentation to the British Association, it seems that Kelly got his share of derision at the end. It is quite obvious though, that neither Kelly nor Bessemer totally understood the mechanisms of the process, although Bessemer was likely far better acquainted with the subject. Kelly was not a good businessman, either, and in 1863 the majority of the rights to the patent were sold to a group that included the Cambria Iron Company, E. B. Ward of Detroit, Z. S. Durfee of Massachusetts, John Morrell of Johnstown, William Lyon and James Park Jr. of Pittsburgh, among others, who formed the Kelly Pneumatic Process Company. The new group successfully dispatched Durfee to England to obtain the American rights to Mushet's recarburization patent for them.[17]

In 1862 Alexander Holley, a Connecticut native, purchased the American rights to Bessemer's patents for himself, as well as John Griswold and John Winslow, both of Troy, New York. Holley was an innovative engineer and rapidly became recognized as the authority on the Bessemer process in the United States. At this early juncture in the history of the process, "The converting vessels [in England] will blow about 250 tons of metal without relining; the [tuyeres] have to be renewed after about 10 tons of steel have been made."[18] The Bessemer process was highly exothermic, producing the highest temperatures recognized to that time. The greatest heat was generated at the tuyeres, where the exiting air contacted the molten metal and began burning the metalloids (silicon, carbon, etc.) that were present, and which were already elevated to a temperature of a few thousand degrees. This was extremely corrosive to the refractory in the immediate area. As noted above, the tuyeres had to be renewed after about ten tons were produced, and with a "charge varying from 3 to 4 tons"[19]; this took only three or four cycles. To make a heat of Bessemer steel in England at that time (c. 1864), the vessel was turned up to a vertical position with the opening at the top, and air was blown through the molten bath for twelve to twenty minutes, after which the converter was turned down to horizontal and the molten spiegel was added. The vessel was then turned back to the upright position and was blown for about five more minutes, supposedly mixing the spiegel with the refined iron. In reality this was not necessary. It was again turned down to pour, and the contents, now steel and not iron, emptied into a ladle.[20] At three to five tons per heat and about thirty minutes per blow, they were capable of making six five-ton heats per twenty-four-hour day in England, although it is doubtful that they operated this way. Since they would have been required to change the tuyeres every four to six heats,[21] producing that many heats in a day meant that it was necessary to make the change at least once and often twice in one day. Replacing converter tuyeres at this time was one of the many horrid jobs connected with working in the steel industry. The tuyeres were extracted with relative ease by removing the bottom plate from the wind box and then digging out the offending item(s) from the back. A new one or sometimes a complete set was inserted in its place. The difficulty arose when sealing the new refractory to the old. Only minutes before the inside of the vessel was glowing with a brilliant white heat at a temperature more than three thousand degrees. To cool down the vessel enough to send a man in there took at least twenty hours, and even then it would not have been a pleasant experience. So they took to pouring a very soupy liquid refractory down through a long pipe suspended from the opening in the top of the converter to seal the void between the old and new. (This opening was called the *nose* because it often looked like one. In later designs, due to a change in the shape to a simple, often concentric, but sometimes eccentric, opening that was flush with the vessel, it was often called the *mouth*.) The new joint was not always sound. The water contained in the soupy liquid rapidly boiled in the intense heat, creating thousands of tiny bubbles that froze in place as the refractory solidified. That made the joint porous and weak, a perfect avenue through which three to five tons of molten metal could leak.[22] Also, the vessel had to be turned up to pour the seal. Working in this way was comparable to sticking one's face above an open fire. Given the circumstances a fire may have been the preferable choice.

The first Bessemer steel (as it is most often described) in the United States was made in late 1864, under the Kelley patent, by E. B. Ward at an experimental works built in Wyandotte, Michigan, about ten miles south of Detroit. The vessel and supports were clearly a close copy of Bessemer's converter. The two-ton plant was built on the grounds of the Eureka Blast Furnace. The iron was run in the molten state directly from the furnace to the converter, where it was first weighed and then hoisted and charged into the vessel. The original cast ingots were sent to the North Chicago Rolling Mill Company in Chicago, where they were rolled into the first steel rails made in the United States.[23]

An unusual accident that occurred at this plant demonstrated the dangers inherent with having molten metal in close proximity to water.

> I first learned that my assistant [William Durfee, who was Z. S.'s cousin and the plant superintendent[24]] . . . having been unfortunate enough to have the tap hole of the casting ladle chill after successfully teeming two ingots, he ordered the ladle emptied into the well, which he had neglected to fill with water; and the result of turning two tons of fluid steel upon about a barrel of water which chanced to be in the well, was a terrible explosion, the metal flying in all directions. Senator Chandler [of Michigan] was prostrated at full length in the pig-bed; Senator Wade [of Ohio] was projected upon a pile of sand in a corner of the casting-house; others of the party were more or less burned, and otherwise injured, while Capt. Ward himself was blown bodily through the open doors of the building into the yard upon a pile of pig iron. For a time everything was in confusion, but it was soon ascertained that by great fortune no one was seriously hurt, and they all returned to Detroit thoroughly of the opinion that they did not care to see steel made by the "new process" again.[25]

This incident dramatically exhibits the fine distinction, both with semantics and fact, between pouring even a relatively small amount of molten metal "onto" rather than "into" standing water. This plant was abandoned in 1869.

In February 1865, Alexander Holley finished the second mass-produced steel plant in this country at Troy, New York, for the firm owned by Winslow, Griswold, and Holley. It was the first to operate under Bessemer's patents. Strangely enough the location for this new converter facility was selected on the basis of an existing water wheel that had been used to power a gristmill formerly situated on the property. The wheel was reused to propel the two pistons of the blowing machinery for the converter. It must have been quite a contrast to see one of the world's most modern metallurgical facilities being driven by one of its most ancient types of power.

The closest supply of iron was some fifty miles distant. The single converter had two-and-a-half-tons capacity. Almost from the start, using the water wheel was a mistake. In May the shaft of the wheel was found to be rotten and had to be replaced, and by December "constant trouble from low blast pressure led about this time to the abandonment of the water wheel." Work was curtailed at the plant until March 1866, when a steam-engine-driven blast was begun. There was trouble maintaining the proper blast pressure from this engine in the beginning. In May of 1866, eighty blows were made for 118 tons of steel ingots with an approximate 20 percent loss in yield. The two-ton capacity plant was expanded by the erection of a second facility that included two five-ton vessels. At the start of 1867 the new works were completed and were "expected to produce 20 to 30 tons . . . every turn of ten hours," or the equivalent of four to six blows. At first the iron for the converters was remelted in a reverberatory furnace, but Holley later changed to a cupola type.[26] In 1868 the plant was severely damaged by a fire, and production was delayed again until it was rebuilt. It resumed operation in January of 1870.[27] The two-ton plant was soon abandoned.

At first, the concept of this seemingly minor shift from using reverberatory furnaces to relying on cupolas for melting pig for the converter might appear to be quite innocuous, but in actuality it deserves considerable attention. The concept is attributed to John Pearse, an associate of Holley who was later the manager of several Bessemer plants. Cupolas were popular sources for hot metal in foundries but were initially found to fail when applied to the steel plant because foundries were only using small lots (about one ton) at the time. It was not until some minor modifications were made to the furnace, such as extending the shaft height in the region between the tuyeres and the tap hole from a nominal one foot to about three or four feet and also by dramatically increasing the cross-sectional area of the tuyeres themselves, did their use become practical. The benefits were numerous. These changes affected the capacity. By increasing the volume of the furnace between the tap hole and tuyeres they increased the ability to hold molten metal. With more air, more coal could be burned more quickly, improving the speed of operation. Cupolas were cylindrical (sometimes elliptical), open, refractory-lined, plate-metal shafts that were very simple. "The cupola is cheaper to construct and maintain, and in it, from five to six pounds of iron are melted with one pound of coal. The cupola melts more rapidly also, and with greater heat, because the coal and the iron are in direct contact, whereas in the reverberatory furnace the melting is accomplished by flame alone."[28] They occupied less room and required simpler surroundings, and one furnace could melt about one hundred thousand pounds of iron in nine hours.

Later the use of coal was abandoned in favor of coke. Less coke was burned in comparison to coal, in the range of one ton of coke to fifteen or twenty tons of iron.[29] Coke burned hotter than coal and was therefore faster as well. The furnace bottoms were frequently set about twenty-five feet or so above floor level, so that the iron could be directed by steep runners to the converter without the likelihood of plugging.[30] Cupolas proved to be a fast, reliable source of hot metal. Their arrival came in time to meet the ever-increasing demands of the expanding Bessemer steel industry. Reverberatory furnaces continued to be used for some time to melt spiegel or ferromanganese due to the potential for sulfur contamination from coal or coke, but even these were converted to smaller cupolas.

Due to the combination of patent control and ownership by Kelly, Bessemer, the Kelly Pneumatic Process Company, and the American Bessemer holdings of Holley, Griswold, and Winslow, for some time it was utterly impossible to manufacture mass-produced steel in the United States without one company in some way infringing upon the patent rights of the other. For instance, the Kelly Company could blow air through molten iron to make steel while violating Bessemer's patent for the vessel on axles and its drive. Similarly, Holley could legitimately use this same equipment but could not legally recarburize it with spiegeleisen or other equivalent material as devised by Mushet, nor could he blow air through the molten metal because the Kelly Company owned these patents. This stymied the development and use of mass-produced steelmaking techniques in America until a suitable solution to this quandary could be found. It was during the construction of the five-ton plant at Troy that the issue was resolved. In 1866 the Kelly, Mushet, and Bessemer patents were all vested in a joint stock company called the Pneumatic Steel Association. One of the important aspects of this conglomeration was a large reduction in royalties and fees charged to those licensing the technology, which helped accelerate the establishment of the Bessemer process in the United States.[31] The stated royalties and charges by the Pneumatic Steel Association were: £1 sterling per ton (2,240 pounds) for ingots for rails; £2 per ton for axles, tires, forgings, and more; £3 per ton for anchors and parts; £5 per ton on ingots for cannons, mortars, guns, and more, and $5,000 for the license to cover costs of information and drawings. The reason royalties were quoted in sterling was to reconcile the charges with those being demanded by Bessemer in England.[32] Later, the ownership of the patents held by the Pneumatic Steel Association of 1866, a joint stock company established in New York, were put under the administration of the Bessemer Steel Company, Limited, which was organized in Pennsylvania in 1877.[33]

The third Bessemer plant in America, the Pennsylvania Steel Company built with two five-ton converters, was completed in 1867 under the auspices of the new Pneumatic Steel Association at what is now Steelton, Pennsylvania, located just outside the state capital in Harrisburg. The impetus for building this plant came principally from the Pennsylvania Railroad (PRR) and its president, J. Edgar Thomson.[34] Holley engineered this facility, and he left his position at Troy to complete the work. It is considered by many to be the first commercially successful Bessemer plant in the country. Others feel this was true of the Troy Works. One important reason for favoring the Steelton plant is that its principal product was rail, and the first commercial order for rail in the United States was shipped via the PRR from that plant as ingots. They were rolled at the Cambria Iron Company in Johnstown (also on the Pennsylvania Railroad) until the rail mill at Harrisburg was completed. In September 1867 the Spuyten Duyvil Rolling Mill Company in New York rolled ingots from the Troy works into rails, but this was only an experiment and not their entry into routine production.[35] Part of the confusion in reliably identifying the first commercial plant arises from the fire at Troy in 1868 that partially destroyed the five-ton plant and stopped production there. This setback put both Troy and the Pennsylvania Steel Company on an even keel.

The move from wrought iron to steel rail was not made for frivolous reasons because steel was significantly more costly than the iron. "The first steel rail was made by R. Mushet [in England] in one of his earliest experiments in 1856, and was laid at Derby Station, with the result that it remained as perfect as ever after six years' wear, though it was in a position in which an iron rail required to be replaced every three months."[36] Wrought iron rails were far too soft and wore out especially quickly at stations or in sharp curves. The rails wore rapidly in curves because of the scouring action of the wheels pressing against the inside of the railhead. In stations, wear was due to abrasion from the locomotive and car wheels sliding across the rails as the train braked to a stop, or when the wheels of the locomotive would sometimes spin as the engineer tried to start the train. Cast iron rails were too hard and brittle, which caused them to break unexpectedly and too easily, especially in winter. Steel was the obvious answer, albeit an expensive one. The average price of steel rails was about $110 per ton in 1872 to 1874. As early as 1863 the PRR tried using crucible steel rails purchased from England for $150 per ton but found them to be too hard.[37] Later they bought softer (lower carbon) crucible rail, and eventually got Bessemer rails from England as well.[38]

The Pennsylvania Steel Company was conveniently located along the main line tracks of the Pennsylvania Railroad. With that road intimately involved in the financing and organization of a company located on its right-of-way, and at the same time desiring the output from that facility, this was a very cozy combination. Still, the railroad was not interested in being involved in the startup of a steel business that was itself technically challenging to say the least. It was already saddled with operating perhaps the most important rail line in the country, and it was beset with many problems arising from its own evolving industry. The main interest was getting the steel rails, which until then were obtained from England at high prices, with availability set by the whims of foreign steelmakers. "When the Harrisburg works started, 3 blows per turn of 12 hours was a full day's work; in 1868 they produced 8 blows per half day, at that time the best practice in the world."[39] With the advent of the Troy and Harrisburg plants, as well as others that would come on line in the not-too-distant future, Pittsburgh was relegated to second-class status in the steel industry, except where it related to high-quality crucible steel.

Holley recognized that a major flaw in producing steel with Bessemer's machinery involved replacing damaged tuyeres. He devised a changeable (often referred to as a moveable) bottom for the converter that he introduced at Troy in 1869.[40] It was premade and dried so that an entire unit could be changed quickly. Bessemer had also devised a way to change a complete set of tuyeres, but his idea still required liquid refractory. With Holley's method, the gap between the old and new refractory was made from the outside at the bottom of the vessel without much exposure to the intense heat encountered using other ways. The seal between the old and the new was made with a pliable refractory material that had a mudlike consistency, which was used to fill the annular gap between vessel and the bottom assembly containing the tuyeres. It wasn't susceptible to boiling, as was the liquid form, and it remained plastic long enough to be troweled quickly and packed into place, eliminating voids and making a solid joint. Originally, it took about two hours to change a bottom of the Holley type. After modifications were made to that design and ancillary equipment such as a bottom car, drying oven, and hydraulic hoist below the converter to lower the old bottom and raise the new one into place were added[41] (in addition to the experience gained from repetitively completing the task), the time required to complete a bottom change was reduced eventually to only fifteen or twenty minutes. The Holley bottom (as it was most frequently called) had an enormous impact on converter output, as well as worker comfort and safety. The idea was patented in 1870 and again for a modified version in

1872.[42] It singlehandedly made possible much of the tremendous increase in tonnage that poured forth from new American steel plants.

> Notwithstanding all the improvements which have been made in the American Bessemer plant, until a late date the best practice only produced eight heats per day with two five-ton converters. The reason of this was chiefly owing to the extreme difficulty, in fact, the impossibility, as it seemed, with the refractory materials at hand, of rapidly replacing the bottoms of the converting vessels when the tuyeres were worn too short for further use . . . As a result [of the invention of the Holley bottom], from sixteen to twenty-five five-ton heats are made per day (twenty-four hours) in the American works. Above 2,000 tons of ingots have been produced at the Troy works, out of a single pair of five-ton converters, and this with tuyeres which stood from five to six heats, and with lining material inferior to English ganister. One thousand tons a month was considered a remarkable production before this improvement was developed.[43]

Another simple, yet extremely noteworthy, development by Holley that was also patented by him involved the layout of the converter building. It was so significant that this design became identified as the American practice versus British practice for Bessemer's layout. In order to maintain the continuity of production, Bessemer steel plants were being built with two and later more converters. One furnace was often out of service for tuyeres or bottom changes, vessel lining repairs or replacement, descaling (removing) of metal buildup at the mouth of the converter, mechanical repairs, and more, making the additional vessel available to continue operations during those times. Bessemer's plan was to set each converter on a low pedestal directly opposite the other around the perimeter of an eight-to-nine-foot deep and roughly forty-foot diameter circular casting pit, whose floor was about eleven to twelve feet below the bottom of the converter. There was a central hydraulic ladle crane that rotated about the pit. Working in the pit could be a living nightmare. The almost incessant cascade of sparks from the blowing operations would fall into the pit where workers were in the process of preparing the moulds, teeming molten steel into them, or stripping the moulds from the solidified ingots. Likewise, splashing molten steel and slag focused by the pit's circular shape would ricochet around the pit searching for men to burn. This was especially true when pouring a converter into the ladle. In warm weather the pit was simply abominable. Hot, humid air heavily laden with smoke and sulfur fumes could not escape this deep depression, insulated by the thick, dirt walls. A breeze was nonexistent, and the glowing red sentinels of ingots lining the arc of the pit not occupied by the converters reflected their immense heat back onto the men. There was little room to move about or flee without encountering other dangers. It was sometimes overpowering to the men undertaking strenuous tasks for long periods, compounding other hazards. Holley's idea was pure genius. He simply placed the converters side by side in a line tangent to the pit and raised them about nine to ten feet in the air. He set them on beams supported by pedestals, providing room underneath to work on the bottom changes, which was something of a challenge with the British layout because they were too low to the ground. Using the beams also opened the area to allow railcars to bring in new and remove old bottoms. The pit was now typically about three feet or less deep. Sometimes the walls were not even a complete circle, being open here and there and allowing for cross ventilation and some breeze, however slight. There were also places for the men to escape even if for only short periods of time, a vast improvement. It also reduced incidences of trapping sparks and splashes of molten metal. While there were still many dangers, they were very much reduced. With the converters side by side there was more room for casting (teeming) ingots, and an additional ingot crane making three instead of the usual two addressed another inefficiency. Also, the hydraulic cylinder shafts for the cranes (both ladle and ingot) ran to the roof, where they were easily supported by transferring the load to the roof trusses, making them much less expensive to build when compared to the heavy counterbalanced British hydraulic cranes (see figure 5.1).[44] Raising the height of the converters did not come without cost. Since the vessels were at a higher elevation, all of the material (coal, coke,

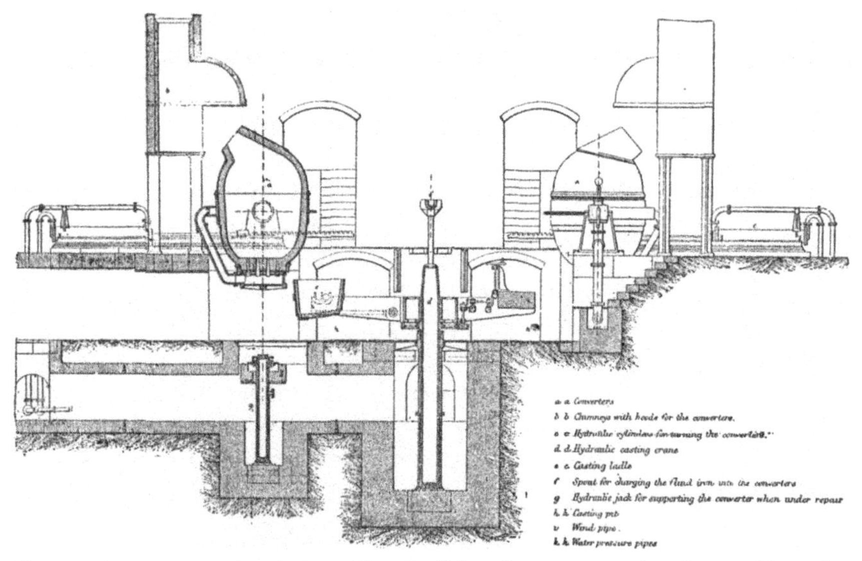

Figure 5.1. Cross sectional view of an English style Bessemer plant layout. Note that the converter vessels are mounted on a tangent directly opposite each other across a deep circular depression. Central to the pit is a heavy, expensive to produce, counter-balanced, hydraulic ladle crane.

Metallurgy of Steel, Harbord.

pig iron, limestone, spiegel, etc.) for the cupolas needed to be raised that much more. Still, even this had a benefit, because the molten iron and spiegel could then be moved from the cupola to the furnace by ladles mounted on rail cars set up on the platforms that were positioned between the vessels to facilitate charging both iron and spiegel, thus eliminating the need for a crane or a hoist. Of course, all of this required additional capital. This was not the first time that vessels were set side by side, but the combination of changes, including the raising of the vessels, providing room for bottom changes, the development of ancillary equipment, and creating more room in a more shallow pit with its easier access for teeming and stripping ingots seems to be the key to its success.[45]

Based on estimates from the Pneumatic Steel Association circa 1869, a pair of converters in a three-ton plant would produce twelve to eighteen tons, and in a five-ton plant from twenty to thirty tons of ingots every ten-hour turn.[46] As reported in 1874, however, Troy was producing an average of 1,700 tons a month, sometimes exceeding two thousand tons, or about sixty tons a day average, from their five-ton plant. Pennsylvania Steel usually made about eighty tons a day but often produced one hundred tons. The new Union Rolling Mill plant in South Chicago, equipped with five-ton converters, averaged about the same as the Pennsylvania plant. The Cleveland Rolling Mill Company's output was 120 tons per day, but they were outfitted with four five-ton vessels.[47] So it can be seen that output was rising. It often took eleven turns a week, and a good estimate is that it required about fifty to sixty men each turn to produce this output from a pair of five-ton converters. The average compensation per man per day was about two dollars.[48]

The output from both puddling and crucible steelmaking was measured in pounds per heat, but a Bessemer's make was tallied in tons. The blowing cycle for a converter took only minutes, while making cast steel and wrought iron required hours. A puddler's job demanded skill to recognize the meaning of the subtle changes presented to him by the appearance of the iron in a furnace bath. He likewise had to possess great strength and stamina to complete his tasks. Similar traits were necessary for a melter making crucible steel. In contrast, a Bessemer operator needed little skill to recognize that the opaque thirty to forty-foot-long ruddy red flame, dense smoke, and accompanying fountain of hundreds of thousands of tiny sparkling meteorites of burning iron, erupting from the glowing mouth of the converter, had quickly transformed itself into a seemingly docile, nearly invisible, transparent white light that announced to him that the steel was done and it was time to turn down the vessel. He did not need to understand that during each stage the color and clarity exhibited by the flame meant a different chemical was being consumed in the inferno, nor did he need to know why. Likewise, he required little strength and no stamina to move the levers that set the converter into motion. To knowledgeable men, these facts were like a fanfare, triumphantly proclaiming to all of the world that labor's stranglehold on the steel industry had ended, that is except where it applied to making top-quality crucible steel.

After the Bessemer steel industry was established, manufacture of huge quantities of acceptable grades of pneumatically produced steel (in relation to crucible steel output) was possible, using workers possessing minimal or no technical skills. This advancement into the manufacture of mass-produced steel and the contrast with its low-quantity, high-quality counterpart would eventually manifest itself as the concept of "big steel" and "little steel."

THE RISE OF THE EDGAR THOMSON STEEL COMPANY

Andrew Carnegie's Edgar Thomson Steel Works in Braddock (formerly Bessemer or Bessemer Station) was built on the site of a historic battle and where the defeat of the British occurred during the French and Indian War in 1755. The fighting resulted in the death of their British commanding officer, General Edward Braddock, for whom the town is named. George Washington was a volunteer officer who fought and served there with Braddock. Many people believed this region was a part of Virginia until later settled by the Mason-Dixon survey. The plant began operation in 1875 and is widely recognized as Pittsburgh's first steel mill. It has already been demonstrated that steel was being manufactured in the Pittsburgh area for more than a few decades, so something must be awry. With the invention of the Bessemer process, where steel could be made quickly and cheaply in large quantities, these new types of plants more or less superseded the previous idea of what a steel mill was. The plant was not a pioneering venture, either. It was the eleventh Bessemer facility in the United States and began operations nearly a decade after pneumatic steelmaking was demonstrated to be a commercial success in this country.

An assertion made by many historians that the reason for Carnegie's naming of the Edgar Thomson Works for the president of the Pennsylvania Railroad was due to his greed. They suggest that it was a plot to obtain orders for rails from that railroad. Rails were the principal (and virtually the only) product of the plant. While I am quite sure that Mr. Carnegie would not have refused such orders from the PRR, I believe that you will find scant evidence to support these implications. First, Carnegie was a former employee of the Pennsylvania Railroad. He had been a subordinate of J. Edgar Thomson during his tenure there. He was also his business partner; and finally, he was a friend. What sort of entreaty Carnegie made to Thomson I do not know, but the reply came in the form of a short, handwritten note from which this excerpt (of a transcription) was taken:

Nov. 14, 1872

Dear Sir—
As regards to the Steel Works, you can use the name you suggest, if the names you sent me are individually liable for its success and as I have no doubt will look after its management—I have no funds at present to invest, having been drained by the Texas and California—and Sanborne's Mexican project—But I can give you Keokuk & H—Bridge Bonds if you can make use of them—

Yours truly

J. Edgar Thomson

To A. Carnegie Esq[49]

Carnegie's solicitation appears to have been more an appeal to a friend to enter a business arrangement than a ploy to gain access to orders from the railroad. The ET Works, as it is usually called, did not begin operations until August 1875. Although construction was started early in 1873, its completion was delayed due to a business depression. Note that the appeal from Carnegie was made in 1872, determined by the date on Thomson's reply, before the delay due to the depression. Edgar Thomson died on May 27, 1874. Both of these instances occurred long before the steel plant began operation in late 1875. If the basis for Carnegie's naming of the plant was an effort to finagle orders from the railroad, then it should have been named the Thomas Scott Works for Thomson's successor and, as will be seen, that would not have been advisable.

Carnegie's first important job was telegraph messenger boy. Eventually, through his own initiative, he learned to read telegraph code and rose to the position of telegrapher at the Pittsburgh office of the Atlantic and Ohio Telegraph Company.[50] Because of his determination to do well, he tried to excel at both of these positions, and he did. Carnegie became so capable as a telegrapher that he was able to read the message directly from the wire, a difficult task. Normally, telegraphers would copy the code and translate later. As luck would have it, construction of the Pennsylvania Railroad between Harrisburg and Pittsburgh was being completed at this time. A frequent visitor to the telegraph office was Thomas Scott, the superintendent of the western division of the railroad and the aforementioned successor to J. Edgar Thomson as president. Scott recognized both Carnegie's abilities and the fact that with all of the attendant activity associated with the construction and operation of a new railroad, he needed a telegraph office of his own. After receiving Thomson's approval, he hired Carnegie as his telegrapher and personal secretary for his office in Pittsburgh. It was Scott who introduced Carnegie to the stock market and the intricacies of investment, much of which today would be recognized as insider trading or worse. Scott even loaned him the money he needed to buy his first stock. This was also Carnegie's introduction to the world of business, right at the heady levels near the top of what was becoming one of early America's foremost companies. Soon after, Carnegie became known to the higher-ups of the Pennsylvania Railroad as "Mr. Scott's 'Andy.'"[51]

Later in life, Carnegie would severely disappoint and then leave Scott vulnerable and unsupported in his investment in the Texas & Pacific Railroad. Both Scott and Jay Gould speculated in the road and expected Carnegie to do likewise. Scott also had Thomson involved in the scheme (as referred to in his note to Carnegie printed above) and committed Carnegie to a $250,000 share for which Carnegie had to and did pay cash to stay unencumbered. Unfortunately for Scott, Carnegie was earnestly preparing to enter the steel business since he had recently observed Bessemer operations in England. When the "business panic of 1873 hit, Scott and his associates found themselves in a desperate situation."[52] Carnegie was in the middle of securing sufficient funds to erect the ET Works at this exact time and was also worrying about the potential effects that the depression could have on his new business. Denying his endorsement of Scott's notes, beyond the already-paid-for $250,000 commitment, was a difficult and painful action for Carnegie to follow (Scott even had Thomson pressure him), but if he had succumbed to Scott, it most likely would have resulted in his personal failure as well. Carnegie's loss of Scott as a friend and the animosity that he later felt severely affected him for the rest of his life. Yet Carnegie always opposed and would never have agreed to use his company to endorse anything that he believed would not benefit the steel company's business.[53]

As a response to this Panic in 1874, through an action of the legislature, the state of Pennsylvania authorized the establishment of limited liability companies. By the end of that same year, Carnegie, McCandless and Company, as it was known at the time, was disbanded, and the Edgar Thomson Steel Company, Limited, was established.[54] Although Carnegie himself was a speculator early in business life, about the time that he established his steel company he changed his view on the subject. Apparently, he was so deeply affected by his experiences with Scott and Kloman that from then onward he looked upon stock trading and manipulating with derision.[55] If Carnegie referred to someone as a speculator, it was likely the lowest opinion held by him.

The Edgar Thomson Works was engineered by and built under the supervision of Alexander Holley. Carnegie offered him the job of managing the plant when it was finished, but Holley declined. He was personally more suited as an engineer, designer, and promoter of steel rather than as a facility manager. Holley had designed nine of the eleven Bessemer steel plants built in the United States to that point. The Kelly plant at Wyandotte and the Freedom Iron and Steel Works located in Lewistown, Pennsylvania, where, ironically, Carnegie is believed to have held a stake in its operation, were the only two plants not built by Holley, and both of them had already been closed and abandoned. Carnegie was not only extremely fortunate in finding extraordinary leaders at what seemed to be critical crossroads for his organization, but he also demonstrated skill in selecting the men for his steel and iron plants. At these ego levels there were many clashes and early exits. Fortunately for Carnegie, while he was conducting his search for an operations manager for ET, there was an upheaval occurring at the Cambria Iron Works in Johnstown, located about sixty miles to the east of Pittsburgh. This was one of America's premier steelmaking operations at that time.

William R. Jones, born in Catasauqua, Pennsylvania, was the son of Welsh immigrant parents. He had little formal education and, due to his father's poor health, was apprenticed at age ten into the service of the Crane Iron Works there. Later, he worked at the Cambria Iron Works in Johnstown under the tutelage of George Fritz. George was the younger brother of John Fritz, who is credited with the invention of one of the more important advancements in iron and steelmaking, the three-high rolling mill. Fritz did not so much invent the three-high mill as made it work. John had worked at Cambria but left to pursue interests at the Bethlehem Iron Company. Jones was present at Cambria while Alexander Holley designed, built, and began operations of the Bessemer steel works there. So Jones was educated in the iron and steel business by some of the most important figures in the history of the industry.

George Fritz died unexpectedly in August 1873. Jones expected to replace Fritz but was passed over by the manger at Johnstown, John Morrell. Morrell, due to his leadership of the noteworthy Cambria facility and his ownership stake in the Pneumatic Steel Association, was an important figure in his own right in industrial history. He did not care for some of the methods Jones used to deal with the men, nor did he approve of Carnegie, whom he is said to have told, "You are too flighty."[56] The result was that Jones accepted an offer from Carnegie at ET where again he would work with Holley, building and starting a new Bessemer steel plant. As an added bonus to his new company, he brought along the superintendents of the rail mill, converting works, machinery, and transportation, together with the head furnace builder and the chief clerk.[57] Later, ET Works employed many other skilled and unskilled workers from Cambria. To say that Bill Jones was well liked by the men who worked for him is a gross understatement.

Bill Jones, frequently called Captain Jones, should not be confused with another important Pittsburgher from the same era, B. F. (for Benjamin Franklin) Jones. B. F. Jones was the "grand old man" of the Pittsburgh iron industry, the head of Jones and Laughlin Company and their American Iron (later Steel) Works that was located several miles downstream from ET on the Monongahela River. Captain Jones's title came as a result of two separate enlistments during the Civil War. He was injured in the course of his first enlistment while crossing the Rapidan. He fought at both Fredericksburg and Chancellorsville. He entered the service as a private but achieved the rank of captain before he was discharged at the end of his second enlistment.[58] It may seem somewhat out of character that a war hero would willingly serve Carnegie. Historians often note that Carnegie paid a replacement to cover his obligation for military service in that war. This was a common and accepted practice during the Civil War,[59] but the implication is that Carnegie was a coward. While it is true that he bought his way out of serving, what most of these same historians fail to record is that early in the war Carnegie, along with Tom Scott, Herman Haupt, and others, were "on loan" from the Pennsylvania Railroad to operate the government's U.S. military railroad. Carnegie was needed for his telegraphy experience and organizational and operational skills. He acted as the superintendent of the military railroad and telegraph offices under Tom Scott, who was the Assistant Secretary of War.[60] It was Carnegie who ordered his subordinates to send a telegraph message to Washington, D.C., informing them that the Confederates had routed the Union forces at the first battle of Bull Run. He could do this because he was there, at Burke Station. Despite the fleeing soldiers, he continued to direct the railroad in removing the wounded back to Washington. Carnegie suffered from heat stroke that day in the hot Virginia countryside. Although he was only a young man at the time, this seemingly temporary disorder ended up affecting him for the remainder of his life. Carnegie was a man of very short stature, and if you care to speculate on what might have been had he served directly for the military, it probably would have been in a staff position in Washington exactly as he had done for the military railroad and not as a frontline solider.[61]

In a letter to Carnegie during the second quarter of 1874, Holley describes the Edgar Thomson Steel Works as being about two-thirds complete. The buildings were fireproof, and the roofs contained no wood.[62] Carnegie hated paying insurance premiums for which there was no financial return. He preferred instead to increase his investment in the physical plant, and once assured that they were safe from fire, he cancelled his insurance.[63] There had already been at least two disastrous fires that destroyed both the Troy (Bessemer) and the Cambria (rolling mill) plants.[64] He preferred that his plant not follow suit. The facility was capitalized at $1 million and consisted of the following: "a 5-ton Bessemer plant [2 vessels]; a Rolling Mill containing a rail-train and a blooming train of the largest capacity, and a full compliment of Siemens furnaces; a gas-producer department; a Boiler department; a pair of 5 ton Open Hearth Furnaces and fixtures; a first class machine and smith shop, with office and store rooms; material sheds, elevated railway over boilers and producers (to save rehandling coal) and a complete water works."[65] See figure 5.2 for a plan of the facility. Originally, there were no blast furnaces at ET, and the iron was brought in as pigs from Lucy Furnace or other sources. The most startling revelation in Holley's communication is the fact that two acid open hearth furnaces had been built along with the Bessemer converters. By this time in his life Holley was recognized as one of the foremost authorities on the Bessemer process in the world. What is not often emphasized is that he was an ardent supporter and promoter of the open hearth process. Mention of these particular furnaces at Edgar Thomson is extremely rare, and they are never ever included in tallies of open-hearth equipment or in *Directories of American Steelworks* of the time. Other sources of information about their existence are often ambiguous. They were used experimentally and their production costs were high, probably an embarrassment and the reason that they were never mentioned. Still, they were available to make expensive grades of steel, and the experience gained using them was probably invaluable.[66] The importance of this installation will reveal itself during the following decade of the Carnegie Company.

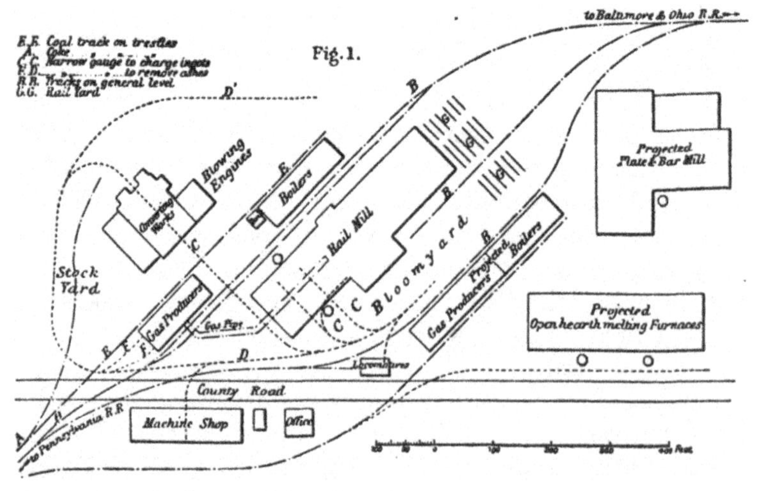

Figure 5.2. Plan layout of the Edgar Thomson Steel Company
From: Engineering, London Vol. 25, 1878.

As occurred with Lucy Furnace, the Edgar Thomson Works rapidly became a legend in its own time, but for making steel. Jones also became a legend. Assuredly, the legends of ET and Jones were synonymous.

> In his first fifteen weeks of steel-making, Jones turned out nearly twice as much as any one had made before with similar equipment.[67]
>
> This company [Edgar Thomson Steel], although its works have been in operation less than four years, reached in 1878 the largest output in rails and billets made by any works in the United States in any one year; and the largest output of Bessemer ingots made by any two-converter plant in the world.[68]

Remember, this was a new company, operating a new process for which it had no extensive historic corporate practice base from which to draw or rely upon other than the former experiences of its employees and supervisors. More important in understanding the extent of their achievement, with the exception of the advancement from smaller five-ton converters to the larger six, seven, and even ten-ton vessels contemplated and later built, most of the major technical improvements to the Bessemer process in the United States, such as the Holley bottom with its associated equipment and the Holley plant layout that vastly increased production potential, had all been implemented long before the Edgar Thomson Works was built. Because the sustained production records from this fabulous plant are fact, it is necessary to conduct a forensic analysis to establish the basis for their success. The expansion of the Bessemer operation at ET was accomplished in stages. In 1878 it was listed as having two six-ton furnaces up from the original five ton, then the move to two seven-ton converters was accomplished by 1880, and a change to three ten-ton vessels was made by 1882, increasing to four of the same by 1890.[69] Also, blowing time was reduced to about eight minutes from the twelve to twenty minutes typical of original English practice. This would seem to be a significant source of increased output. By doubling the vessel size and halving the blowing time, it should have resulted in a fourfold increase in capacity. Yet according to one of America's most respected metallurgists of the time, Henry Marion Howe, "the rapid working which has led to such enormous outputs from American mills is not due to rapid blowing, but to avoiding delays between the blows."[70] Since Howe's work was produced after the amazing increase in U.S. production was already a fait acompli, it is necessary to apply his insight to what must have occurred at ET. Considerable credit must be given to Holley for his magnificent original design of the plant. Yet because he died in 1882 at about age fifty and had been feverishly working on other matters for some period before the time of his death, his credit must be limited only to the very early stages of the facility's operation. Likewise, contrary to what it might infer, expanding from two converters to three did not equate to a time-and-a-half increase in output over the two-vessel operation, because conditions are often far more complicated than a cursory analysis will allow. Sometimes logic does not always work in the way that it seems that it ought.

The changes promulgated by Jones were for the most part not large, at least not in a dramatic sense. They were usually subtle, but numerous. Some of his more important ideas were more esoteric and are mostly intangible. From the beginning, Jones established an esprit de corps among the men at ET, probably one of the most crucial things that he accomplished. The basis for establishing this communion with the men goes back to his working at Cambria and even earlier. At Johnstown, Jones subscribed to the philosophy of George Fritz. When Daniel Morrell, the manager at Johnstown, demanded a reduction in wages as a means of cutting costs, Fritz opposed him, even though he was Morrell's subordinate. Fritz wanted to keep the pay scale high and make the desired gains by increasing productivity. Morrell's view was quite the opposite, but Fritz had the support of sufficient members of the management at Cambria in addition to the backing of the workingmen, and it culminated in his triumph over Morrell. When Fritz suddenly died, it was Morrell who got to choose his successor, and because Jones backed Fritz on the wage issue, he was passed over for the job.[71] So Jones moved to Pittsburgh and battled Carnegie on the wage issue at ET as well. At one time he told him, "Low wages do not always imply cheap labor. Good wages and good workmen I know to be good labor."[72] Jones devised and used a short menu that he believed would guarantee a good workforce:

- Mix the nationalities of the men but steer clear of Englishmen and "westerners"
- Employ men that are young and ambitious
- Keep the works running steadily, permitting a lower wage
- Make the men feel that you are interested in their welfare
- Treat them well and try to make their cost of living as low as possible
- Create a "strong but pleasant rivalry" among the workforce
- Work an eight-hour day
- Use the most up-to-date equipment[73]

The last of these is a reference to the famous "scrap heap" policy at ET and is one of the more famous ideas.

Jones practiced what he preached. He was noted for being a charitable man. He was a tough but fair boss. Of course, that could depend on one's perspective. Jones showed the men that he was concerned about their work conditions. He personally communicated with them and acted to address and correct poor conditions that were within his control, at least within the ideological constraints of the era in which they worked. He demanded that they do their job, respect their workplace, and be responsible in the community. Jones was known to fire people and then immediately hire them back when he realized that he had overreacted. He was also willing to confront, and fire if necessary, men whom he honestly believed wrongfully failed to pay their just debts incurred outside of the plant.[74] "Jones was the only supervisor in the company who could and would serve as labor's spokesman to Carnegie."[75] Of course, one could use all of the same attributes to say that Jones was a spiteful, overbearing monster. I think your judgment would be incorrect.[76]

Unbelievable as it may seem, in 1878 Jones persuaded Carnegie to institute a three-turn, eight-hour workday.[77] He convinced Carnegie that twelve hours of strenuous work, often in appalling heat, was too much for many men to endure. He promised that output would increase, and with the men working four less hours a day with more rest, there would be less waste, lower absenteeism, and fewer accidents. Also, because conditions were improved, a better-skilled workforce could be employed.[78] Using Jones's own words:

> In increasing the output of these works, I soon discovered that it was entirely out of the question to expect human flesh and blood to labor incessantly for twelve hours, and therefore, it was decided to put on three turns, reducing the hours of labor to eight. This has proved to be of immense advantage to both the company and the workmen, the latter now earning more in eight hours than they formerly could in twelve hours, while the men can work harder constantly for eight hours, having sixteen hours for rest.[79]

Regrettably, the rate of return on this was intangible and was negated when H. C. Frick arrived as a partner in 1882.[80] He soon brought pressure on Carnegie to change that policy. The rest of the industry did not rush to the eight-hour day. No one else in the Pittsburgh steel industry adopted it. So with Frick's persuasion, in combination with Carnegie's innate desire to see actual cost figures (savings) on paper, he succumbed to Frick, and in 1888, he forced Jones to return to the two-turn, twelve-hour day.[81] Above all, Frick viewed labor as nothing more than a commodity. It must be added, though, that Carnegie would sometimes seem to succumb to pressure, but in fact it was what he really wanted all along. Unfortunately for Carnegie, this is one of a number of instances in his life where his failure to take a definitive stand on the side of men in a personal conflict pertaining to a humanitarian versus an economic issue reflected very poorly on his character. Wall believed that "to answer these questions requires a far more profound insight into Carnegie's personality than the simple charge of hypocrisy can provide. In part, the answer lies in Carnegie's vanity, in his desire to be loved and admired by all Americans."[82] It would take in excess of thirty more years until the steel industry would finally face up to this issue and dispense with the twelve-hour day for good. By that time Carnegie had been retired from the steel business for more than twenty years, and both he and Frick were dead.

Bill Jones established and fostered a competitive spirit among the men. An astonishing fact is he used a common broom to accomplish this. He found a very large push broom and would have it hoisted above any department that broke or "swept clean" a record. As that record fell, he would have such a broom placed high above the factory area for that group, for all to see. The broom became a badge of honor and pride. Increased productivity, along with some fun and diminished drudgery, were some of the positive spinoffs. Later, records fell so frequently that there wasn't time to hoist the broom.[83] Sometimes Jones, to the consternation of Carnegie and others, played a friendly game of baseball with the men after they had completed their tasks.

Another obscure, yet important aspect, contributing to the rise of the Edgar Thomson Works was the development of the accounting system by William Shinn. He was the first general manager of the plant and Jones's boss. Shinn had been an executive with the Allegheny Valley Railroad, the line that served the Lucy Furnace, which was affiliated with and later owned the Pennsylvania Railroad. Previously, he had been the general bookkeeper for the Pittsburgh, Fort Wayne and Chicago Railway and held executive positions there as well as at other members of the Pennsylvania Railroad's stable of lines.[84] Both the PRR and Baltimore & Ohio Railroad served Edgar Thomson. This was one of the reasons for the steel company's decision to locate its plant in Braddock. By maintaining the competition between the two railroads, the company would not be beholden to whims and control of either one. The voucher system for keeping track of costs that Shinn initiated was used by railroads for many years and by Standard Oil as well, but it was a first in Pittsburgh's iron and steel industry.[85] Before that time, "rule-of-thumb methods prevailed in making iron and steel, and even leading manufacturers did not keep track of the exact cost of each process. At times, when books were balanced and stock was taken at the end of the year, ironmasters who thought they were making profits showed losses and vice versa."[86] Shinn had developed an accounting system for tracking operational and financial details that was partially adopted by both state and interstate commissions.[87] Carnegie, with his experience as a railroader and a manager, would certainly have been familiar with them. From the beginning at ET, Shinn provided Carnegie with a monthly synopsis of the pertinent costs and profits for ET—for example, labor, cost of metal, other materials, selling price of rails, and more.[88] Eventually, this became much more detailed. So much so, in spite of its initial appearance to be unrelated, that the cost reports were eventually used as an investigative tool by Jones to focus on an area highlighted in the report. He would use these costs, together with his mechanical and managerial skills, to pinpoint the root cause of a problem in the unit found to be out of line.[89] The importance of using these accounting methods, as well as the extent to which they were applied, is easy to dismiss based on our modern mind-set accustomed to megasized, computer-oriented accounting departments, but their use in the iron and steel industry was practically unheard of in this era. Because costs were accurately known, the price that could or should be charged for rail could be precisely gauged. And an estimate of the profit to be made at that price, if any, could be known, although, frequently, the price paid for rail was based on what the market would bear regardless of the cost of production or profit/loss. With this information, Carnegie's company could judge what their level of participation in a sale of rails should be better than its competitors. Carnegie was always seeking information about wages, production costs, material costs, labor usage, and more from friends or acquaintances at other steel companies. He was not above providing "gifts" for receiving the information that he used to compare to his own. Even these sometimes-doubtful numbers would be used to try to compel his managers to do better. According to Bridge, a note from Carnegie on the back of one of his business cards would be dispatched to the offending manager(s). In a somewhat humorous vein, Bridge quotes Jones as chiding his managers about one of these notes from Carnegie about costs, by saying,

> "Puppy dog number three . . . you have been beaten by puppy dog number two on fuel. Puppy dog number two, you are higher on labor than puppy dog number one."[90] But from Wall there is better insight: "Captain Jones was a particularly easy victim for Carnegie's constant goading on costs and production, for he carried much of the fierce competitive sportsmanship that he demonstrated in playing baseball, which he dearly loved, into the mill."[91]

One of the more notorious and legendary policies to come out of Edgar Thomson in this era was known to the industry as the "scrap heap," an explanation of which will follow. A much-publicized and often-quoted statement made by Carnegie was: *Pioneering don't pay.*[92] Yet in many respects he and his company became the pioneer, the role model for others in the steel industry.

Jones, like Kloman, was a mechanical genius and was granted at least thirteen patents in his lifetime for ideas ranging from a very simple one, involving armoring applied to hoses, to the highly touted mixer to be discussed later. He once installed a mill at ET with an open-type mill-stand, even though Holley pointed out to him that this same type of installation was a total failure at another plant. Despite the fact that Holley was a trained engineer and he was not, he stood up to the renowned designer and inventor with the response that it was not built with the size and type of bolts that he intended to use.[93] Jones was a complete package: he knew and understood how to make iron; how to make steel; the intricacies of rolling, maintenance, and troubleshooting of the machinery that he operated; and how to motivate men as well. When he told Carnegie that better productivity and better quality with fewer breakdowns could be had by changing out an expensive piece of mill equipment, it would be replaced without hesitation, even though it might cost tens of thousands of dollars or more and the "obsolete" equipment might have been only a few months old. The "out-of-date" machinery was then relegated to the scrap pile or "scrap heap" to the astonishment of the rest of the industry. Evidence of Jones's ingenuity can be found in his work at the rail mill.

> At the stands were six men, three on each side, who with hooks suspended from above caught the rail as it passed through and lifted it to the next pass . . . and there were 12 to 14 rail passes in all . . . One of the improvements that Jones made was the automatic roller table, in 1877, which was operated by a single man and displaced the hook and tong method of handling rails from one pass to the next.[94]

Not only did this save the labor cost of the ten additional men (two turns) formerly used, it also saved the men from exposure to the extreme heat of the red hot rails, the grueling work of lifting them, the potential to be struck as a rail exited the mill when they attempted to catch it, and/or from being killed or maimed if they had the misfortune to be pulled into the rolls. It was probably faster as well. Carnegie always had the accounting system to prove that these ideas were prudent or not. Paying such close attention to detail usually has an immeasurably positive return on profits. Modern safety programs, for example, usually have a large cost and no monetary return associated with sustaining them. Yet companies that promulgate safety and rigorously adhere to these ideals often have good quality, high production, and high profit because of the attention paid to the other aspects of the operation in the same manner—an eye on the very same things, the details.

Carnegie could never corral Jones into becoming a partner, as he had done with so many others. He used partnership as a means of both rewarding and controlling them. Carnegie had such a large share of the holdings in the associated companies that even with the small dividends paid, he had more than enough money to keep him happy. This was not so true for the other partners. Jones was often prepared to resign for one reason or another, and after one confrontation in particular he told Carnegie that he would only stay if he had a sufficiently large salary to make him feel worthwhile. What salary he wanted he would not say, and he left it to Carnegie to come up with a figure for his worth. Jones did inform him that the superintendent at Cambria received $20,000 a year, and he thought that in the range of $15,000 would be right for him, but he would accept less. Carnegie's response was that he would pay Jones the same amount as that received by the president of the United States, which at this time was $25,000 a year.[95] By the end of his tenure not too many years later, Jones was earning $50,000, including royalties. Even at that, Carnegie got a tremendous bargain.

As previously mentioned, the blowing operation for an American Bessemer plant was reduced to about eight minutes in the early part of the Edgar Thomson era. Making a heat of Bessemer steel involved far more than just blowing, and as Howe has already intimated, it was what occurred between the blows that was of crucial importance. The entire operation can be broken down into a relatively few important steps, which can be identified as charging, blowing, tapping, and teeming. These phases are briefly described below and can be further partitioned into a comparatively small number of additional operations:

- Charging involved filling the vessel with five to ten tons (and later more) of molten iron, which was usually brought in via a ladle but sometimes early in the Bessemer steelmaking era it was discharged directly from a cupola. The vessel was laid back to a horizontal position to receive the charge. The air blast was turned off to tap. The converter's shape was designed to permit receiving or discharging the proper quantity of metal without the air being turned on. Lying on its side, the normal level of molten metal was too low to enter the tuyeres or air box while the air was not flowing. A cursory inspection of the furnace's lining and tuyeres was performed before the charge was made to determine if either needed to be repaired or replaced, intending if necessary to change the bottom or patch the lining after the current charge.
- Blowing and tapping. The air was again introduced into the tuyeres, and the vessel was turned up to blow. As previously stated, this operation took very close to eight minutes at the fast mills, after which the vessel was rotated forward or turned down and the air was again shut off. The recarburizer, in the form of spiegel or ferromanganese, was then added to the refined metal in the horizontal vessel. This was usually accompanied by a substantial reaction in the converter, which was then tilted slightly further forward, and the steel was poured or tapped into a waiting ladle for teeming. Once the steel was emptied and the crane moved the ladle away, the converter would then be rotated to the upside-down position to dump the slag onto the floor of the shop for cleanup.
- Teeming's primary function was to fill the ingot moulds with molten steel from the ladle, but in actuality it entailed much more. The teeming opera-

tion was almost totally separated from the converter with the exception of their joint use of the ladle crane. These additional steps were a large portion of what Howe referred to when he wrote of the time "between the blows" and included: waiting an appropriate time for the molten metal to cool in the mould (solidify) sufficiently before it could be removed or *stripped* from the ingots; removing the red-hot ingots from the pit; and setting the new empty moulds into position to receive a new heat of molten steel. This portion of the work required the use of pit cranes, but all of the operations had to occur within the slightly more than eight minutes that was required to charge, blow, and pour a heat from a converter.

This would have been a relatively simple process had there been just one vessel operating, but more typically there was another one already blowing that would need to be tapped within a few minutes, and only one ladle crane was available. Ingot cranes were doing the stripping, removing ingots, and replacing moulds for the next cast, while the ladle crane was returning to get the next heat from the second converter. The ladle required inspection or might need to be replaced before receiving the next heat. Howe made a detailed analysis of well over five hundred cycles of Bessemer operations prior to 1890 with a level of precision that determined the time taken to perform a particular task to a resolution of the number of seconds. To get a broad view of the subject, his data was collected from a number of different plants across the United States. See table 5.1 for an examination of the performance at a ten-ton plant. Obviously, the time required to teem would be dependent on both the size of the converter and the size of the ingots being cast. Much of the time, at least early in this business, 1.25-ton ingots were a norm. This permitted rolling four sixty-pound rails per ingot. The weight of the rail given here refers to a standard that is still in use to this day and is defined as the weight of a three-foot section of that particular profile of rail. Some would call it sixty pounds to the yard. For a five-ton converter there would be four 1.25-ton ingots to teem, for a six-ton converter it would be five ingots, for a seven-ton would be six ingots, and for ten tons there would be eight. The number of ingots cast became a limiting factor as far as converter vessel size was concerned due to the time constraint for teeming. Thus, for a number of years it seemed like ten tons was the upper limit of converter size, although considerably larger vessels were contemplated and later achieved. Some companies, like the plant built at Homestead in the 1880s with two four-ton and later two six-ton vessels, got around this issue by teeming larger ingots. However, what seems a simple solution introduced its own set of issues. At this time, many rolling mills took longer to roll an ingot than mills that were built later, which were faster and more powerful. The larger ingots often cooled too much before rolling was completed, and frequently these rails were not first quality.

A converter bottom usually lasted from fifteen to thirty blows or an average of about twenty-five heats, while the vessel lining typically survived about a week before requiring extensive repairs.[96] This meant that about every four to five hours of normal operation the bottom needed to be changed, but except for occasional minor patching the lining would usually last until the downturn during the weekend. As garnered from the data in table 5.1, lines II and III, changing a bottom took fifteen-and-one-half minutes to accomplish on average (III minus II). Diligent inspection of the converter prior to each charge usually gave fair warning that a required change was imminent, and, "as soon as vessel 1, whose bottom is worn out, finishes its heat, the work of changing bottoms begins; even before the steel is poured into the ladle the bottom is partly unkeyed."[97] Keys were relatively small, wedge-shaped plates that were quickly and easily sledgehammered into or out of place and were used to secure the bottom to the vessel through mating loops attached to both parts. They took the place of nuts and bolts or similar hardware that was easily damaged and were more complicated or demanded considerably longer to install or remove. Jones modified Holley's bottom and key arrangement so that his joint between the vessel and the bottom was at a much shallower angle (flatter) than Holley's (see figures 5.3 and 5.4). From the diagram you can see that the benefits were that the refractory cement for the joint could be applied to the new bottom as it rested on the bottom car at ground level. After the old bottom was removed, when the new one was hoisted up against the vessel with the hydraulic lift, the pressure from the lift squeezed the excess mortar out from the joint between the two without the need to trowel, reducing the time required to complete the change. Preplanning made it possible to make two quick blows on the other vessel with only about three minutes, twenty-five seconds in lost production. The total amount of working time lost for bottom and lining repairs at a two-vessel operation was only 5 percent, so surprisingly, this was all that one could expect to gain by having a three-vessel plant in similar conditions rather than the seemingly logical time-and-one-half output of a two vessel plant.[98] This does not account for the time lost when a vessel had to be completely relined. "In a word, the three-vessel plant works a little more easily and hence a little more cheaply. On the other hand, the three-vessel plant is necessarily a more expensive one."[99]

Most of the heat generated in a Bessemer blow came from the consumption of silicon that was contained within the molten iron. Quick cycling from one heat to the next meant that there was little temperature loss in the vessels between blows. This permitted the use of lower silicon iron due to the heat saved by not letting the vessel cool (or the steel would then be too cold when

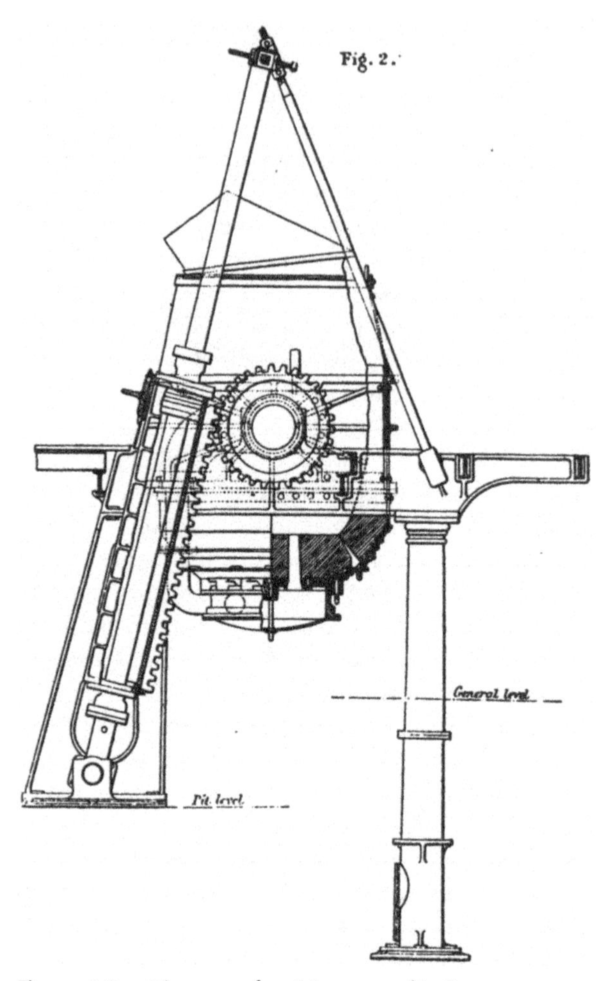

Figure 5.3. Diagram of a 6-ton capacity Bessemer converter of the Holley design used at the Edgar Thomson Works. Note the steep angle of the joint between the Holley design of removable bottom and the vessel proper shown in the cut-away section at the bottom right (the triangular white gap).

From: Engineering, London Vol. 25, 1878.

tapped and might cause a problem with teeming). With the vessel already very hot and with less silicon in the iron, even quicker cycles were permitted because of less silicon that had to be burned out. This reduction in the amount of silicon in the iron quickly reached the law of diminishing returns. Curiously, though, it sometimes got to the point where scrap had to be added to

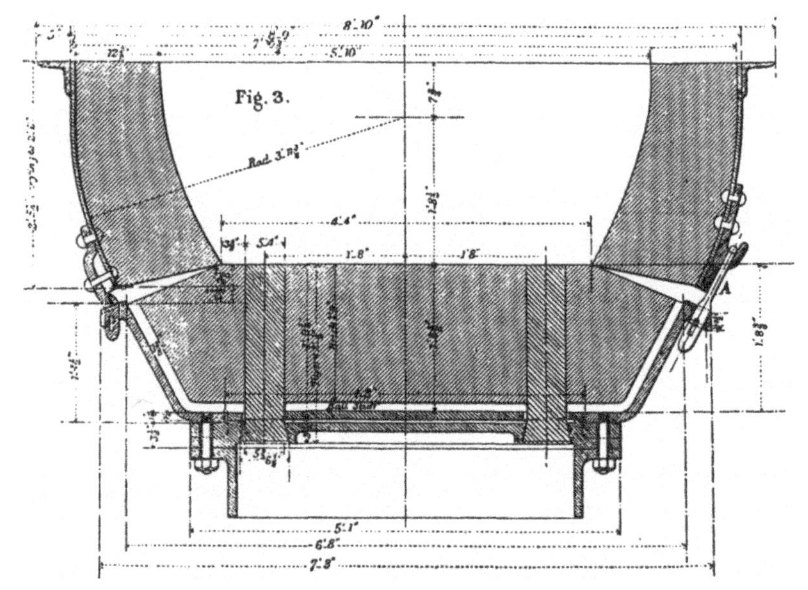

Figure 5.4. Diagram shows Jones' modifications to the Holly bottom. Note the shallower angle (flatter) of the joint between the bottom and the vessel proper when compared to Holley's design in figure 5.3. This simple adaptation facilitated faster sealing of the interface between the two using a pliable refractory to complete the change and resulted in less time required to execute the task. That same purpose was behind the cotter-link system shown on the outside of the converter which mated the two parts together, that is, for quickly detaching the old and then re-securing the new bottom to the vessel by using the links and wedge shaped keys instead of bolts and nuts or similar fasteners.

From: Engineering, London, Vol. 25, 1878.

the vessel during a blow, not as a means of recycling, but for use as a coolant, because the steel was getting too hot, creating its own problems. Adding a coolant required additional time, which decreased productivity. Jones devised a method by which he would introduce steam or even a spray of water into the blowing air to reduce the metal temperature, eliminating the need for *scrapping*, thus saving time.[100] Another method used to shorten blowing time at ET and other rail manufacturers was to stop blowing early; that is, turn down the vessel as soon as the flame intensity began to drop off. Although the amount of time saved was only a matter of seconds, one can begin to understand that saving time, even a short duration, was of utmost importance due to its cumulative effect on the process. The jargon for implementing this procedure was later referred to as *blowing young*, which was possible because of the higher carbon content in rail steel. This ultimately resulted in saving spiegel, and as a bonus it reduced losses from metal being blown out of the converter.[101]

Table 5.1. Time Required to Perform Various Steps in the Bessemer Steelmaking Operation of a Two-Vessel, Ten-Ton Plant

Description of Process Step	*Holley Plants Measured Time*	*Minimum Time Measured*	*Comments—Time Depends Chiefly on:*
a. Charging cast iron and inspection of the tuyeres and lining.	1'-50"	1'-40"	
b. Turning the vessel up and down.	0'-30"	0'-19"	
c. Blowing.	7'-30"	. . .	
d. Recarburizing.	0'-25"	0'-15"	
e. Pouring into the casting ladle.	0'-45"	0'-40"	
f. Emptying slag and returning back to position.	0'-25"	0'-22"	
g. Swinging the casting ladle to the moulds.	0'-28"	0'-25"	
h. Teeming ten tons into eight ingots.	5'-30"	5'-15"	
i. Changing or repairing ladles.	1'-00"	0'-50"	
j. Swinging the crane back to the vessel.	0'-13"	0'-11"	
1. Time each ingot must stay in mould.	10'-20"	9'-49"	
2. Stripping eight ingots and setting their moulds on cars.	3'-30"	3'-00"	
3. Lifting eight ingots and setting on cars.	3'-30"	3'-20"	
I. Total blowing cycle (b+c).	8'-00"	8'-00"	
II. Vessel cycle time between blows without bottom change (a+d+e+f).	3'-25"	2'-57"	
III. Vessel cycle time between blows as above with bottom change time added.	19'-00"	17'-00"	
IV. Cycle time of casting ladle (e+g+h+i+j).	7'-56"	7'-11"	Weight of charge & # of ingots
V. Cycle time of moulds. *	19'-47"	18'-37"	Thickness & # of ingots
VI. Cycle time of ingot cranes. **	10'-10"	9'-20"	# of ingots

* Determining this number was a bit more vague as it required the use of both the casting and the ingot cranes and their respective crews and consisted of the time to teem the ten tons of steel into eight ingots—5'-30—and then, in addition to this, it depended on what occurred after the last ingot was teemed, that is, the last ingot must stay in the mould, 10'-20"; to strip the last ingot took 0'-27"; to lift the last four ingots, 1'-50"; and finally, to replace last four moulds, 1'-40", which totals 19'-47". The first mould was stripped by the ingot crane 4'-50" after the last ingot was teemed and the ladle crane started to swing back to the converter (10'-20" minus 5'-30"), and successive ingots followed.

** This total represents stripping eight ingot moulds (3'-30"), lifting the ingots and putting them on cars (3'-30"), and setting the eight new moulds in place (3'-10") or 10'-10" total.

Source: Henry M. Howe, "Notes on the Bessemer Process," *Transactions of the American Institute of Mining Engineers XIX*, New York, 1891, 322.

The primary limiting factor for continuous production in a two-vessel shop as shown from table 5.1 was the blowing time (I) combined with that of the vessel cycle time without a bottom change (II), or a total of eleven minutes and twenty-five seconds, versus the cycle time of the casting ladle (IV), which was seven minutes and fifty-six seconds, or a positive cushion of about three-and-one-half minutes between them. Under these average conditions a ladle would always be waiting to receive a new heat of steel from a blow within an acceptable margin. That was true as long as the newly filled ingot moulds were stripped, ingots were removed, and fresh moulds were placed and waiting. By examining items V and VI, it can be seen that they met these constraints.

Each step of this process was timed to a resolution of seconds, and each of those seconds was an important element to the total success of the process. Furthermore, each segment was dependent upon the execution of the others from the first to the last. Is this not essentially the concept of the assembly line? Sure, it was not connected by a continuous conveyor, nor was it a highly complex system, but it did consist of an interconnected permanent path of interdependent sets of copulas, vessels, rails, and cranes to manufacture a product from beginning to end. This was practiced decades before Henry Ford and others were credited with its invention.

In the latter 1860s, production of a few hundred tons of ingots in a month was thought to be good work. In the early 1870s, prior to the inception of ET in 1875, 1,500 to 2,000 tons per month was considered to be an excellent output of steel.[102] In February 1876 Edgar Thomson produced 2,240 tons of ingots, or what was probably the largest output from a pair of five-ton converters anywhere in the world.[103] By 1880 the plant was routinely turning out 3,000 to 3,400 tons of ingots, or almost twice the pre-1875 monthly tonnage, but in only *one week*.[104] According to Jones, from a speech he made in Britain:

> The output of the American works is governed by the facilities for getting the ingots on the road. This is the sticking-point just now. Therefore the works that cast their tonnage in the least number of moulds have a decided advantage in reaching the ultimate production of the present American or Holley plant. The race, so far as the Edgar-Thomson Works are concerned, will soon cease. . . . Edgar-Thomson will change from a two 7-ton converter to a three 10-ton plant, and then our efforts will be concentrated upon keeping pace with the Bethlehem four-vessel plant, and with the North Chicago and Pennsylvania Steel Company's three-vessel plants.[105]

Jones exceeded his goal; data from table 5.2 for 1889 indicates that ET was able to produce more than 7,500 tons in a week, or more than double its 1880

Table 5.2. Maximum Output of American Bessemer Works*

	Output in Tons			*Number of Heats*			*Work*	
Plant Name	*Day*	*Week*	*Month*	*Day*	*Week*	*Month*	*Hours per Day*	*Shifts per Week*
Two-Vessel Mills								
Union (Chicago)	. . .	. . .	28,145	. . .	. . .	2,858	12	11
Scranton (PA)	. . .	. . .	. . .	. . .	760	. . .	. . .	. . .
Homestead (PA)—rail steel, 1887	891	4,477	19,572	170	809	3,636	8	16
Homestead (PA)—low carbon steel, 1889	619	. . .	13,291	116	. . .	2,436	8	16
Three-Vessel Mills								
Edgar Thomson	1,384	7,557	31,120	133	719	3,014	12	11.75
Harrisburg (PA)	1,029	5,110	20,947	141	678	2,929	12	11
South Chicago	1,393	6,402	27,427	119	556	2,441	12	11

*Data current to circa May 1889.

Source: Henry M. Howe, "Notes on the Bessemer Process," *Transactions of the American Institute of Mining Engineers* XIX, New York, 1891.

output. This was substantially more than the production at either Chicago or Harrisburg. (See figure 5.5 for a photo of ET's four-vessel plant.) Even after accounting for the greater output gained by increasing the size and/or number of converters at Edgar Thomson, consideration must also be given to the fact that it was operated like a finely tuned machine. Prior to the expansion of the plant, when Jones was still using the old converters, by slightly increasing the level of molten iron in the vessel, he gained additional output. Steel weighs a little less than five hundred pounds per cubic foot, or about one ton per four cubic feet. This volume would be very slightly larger if you use the 2,240 pounds per English ton as a standard. The present-day metric ton is 2,200 pounds or 1,000 kilograms. If you increased the depth of molten iron in the vessel by about one inch in an eight-foot internal diameter converter or by about two inches in one of six-foot diameter, the output would increase by one ton, and this came at the minor cost of increasing the blowing pressure by about a quarter to a half-pound. The depth of molten metal in a converter of this era (c. 1890) ranged from 12.5 inches to 15.5 inches, which represented about seven to ten tons.[106] Manipulating the depth of charge required careful thought and demanded implementation in increments of a complete ingot or an amount that would evenly increase individual ingot sizes so that a complete complement of the larger-sized ingots was made. Otherwise it would just make so much additional scrap. The evidence of Jones's actions of increasing the output of the converters in this way came from the same talk previously mentioned. Jones indicated that the seven-ton plant was pushed to make 7.25 to 7.75 tons (of 2,240 pounds) per heat, facilitating the making of four sixty-seven pound, five sixty-pound, or six fifty-two-pound rails of thirty feet length from five fourteen-inch ingots. This weight of rail would indicate that the ingots were slightly heavier than the 1.25-ton norm expected from five ingots produced by a seven-ton converter.[107] Jones initially preferred rolling single lengths of rail rather than double lengths because it was his belief that fewer second-quality rails were produced.[108] Later, after they had the rail saws installed, they would actually roll four lengths at one time.

> Leadership was a key to the success of ET. As Jones told the association in London: . . . the development of American practice is due to the esprit de corps of the workmen after they get fairly warmed to the work. As long as the record made by the works stands the first, so long are they content to [labor] at a moderate rate, but let it be known that some rival establishment has beaten that record, and then there is no content until the rival's record is eclipsed.[109]

Note that Jones did not take the credit for himself; he gave it to the men. Another indication that points to the successful adaptation of ET to the changing environment is demonstrated in table 5.3. Notice that the selling price of rails steadily declined during 1877, while the cost of production declined as well, so that rails were always at least marginally profitable as far as comparing the figures for the month for which the data was reported is concerned. It must be noted that one cannot match a declining selling price with a decreasing production cost, unless there was an accurate means of tracking the costs of operation.

Howe's much oversimplified, but quite succinct, summation of the reasons for the success of high-volume American Bessemer practice was as follows:

> Short blows imply low silicon and fast blowing. Low silicon implies short intervals and fast blowing. Our blows are short because we have little silicon to remove, and because we remove it fast by supplying blast rapidly. We get enough heat even with little silicon, because our blows and the intervals between them are so short that relatively little heat escapes our vessels . . . Finally, our short intervals imply powerful machinery, efficient organization, and extreme specialization and subdivision of labor, which can be profitable only when the output is large.[110]

The managing director of Bolckow, Vaughan of Middlesbrough admitted at the London meeting that the output of the ET converters was just double what his own company could make from its almost identical plant. There was certainly an impressive work ethic at ET. It should be noted here that English plants did produce some large runs of rail products. However, their country was so much less expansive than America, and most of their rail lines had already been laid, so these companies didn't have the same opportunity as U.S. rail producers. Switching back and forth between product lines and dealing with higher levels of phosphorus and lower grades of iron ore required increased blowing times and raised production costs.[111] British and other European firms were struggling to meet orders of all sizes, and differing specifications won in keen competition with each other.[112]

By the Panic year of 1873, six Bessemer furnace shops were operating in America: Rensselaer in Troy, New York; Cambria in Johnstown, Pennsylvania; Pennsylvania Steel near Harrisburg, Pennsylvania; North Chicago Rolling Mill and Union, which were both in Chicago; and, finally, Newburg in Cleveland, Ohio. By this time two of the original group of eight had already fallen by the wayside: America's first experimental works at Wyandotte, Michigan, and the facility in Lewistown, Pennsylvania. Two more came on line that year to take their places, one at Joliet, Illinois, and the other in Bethlehem, Pennsylvania. Increasing output was met by falling rail prices, dropping from $158 per ton in 1868 to $120 by 1873 (see tables 5.5 and 5.6). After a short respite in construction due to the Panic, two more plants, the eleventh and twelfth, began operations in late 1875, at Carnegie's mill at Braddock and the Lackawanna Iron Company in Scranton, Pennsylvania.[113] Rail prices continued to plummet as the market was deluged by even more production capacity, especially from the miracle at Braddock. The price of rail dropped by more than half from the beginning Panic price of $120 to less than $60 by 1876 (table 5.5).

In 1875 American steel and iron industry manufacturers joined together and formed a new trade organization, the Bessemer Steel Association. In order to stabilize prices in the rail industry, they met the following year to form a cartel, or what was typically known as the *rail pool*, to set the selling price and allocate the market among the member steel companies. Naturally, in accordance with the principles of Orwell's *Animal Farm*, some of the steel company members were more equal than others and therefore got a bigger share of the market. Already noted by many members for his arrogance, Carnegie was in attendance. Those in control of the pool did not consider his new plant in Braddock to be in the upper ranks, so the Edgar Thomson Works was allotted 9 percent of the market, the very lowest level, whereas Cambria got 19 percent and Pennsylvania Steel 15 percent. All of the other shares fell somewhere in between. ET was on the bottom.[114]

I'm sure that some of Carnegie's perceived arrogance came from the fact that he was not "one of the boys." He was living in New York and was not much of a social member in the clique of steel men. Carnegie was not so familiar with the technical operational aspects of making steel, but he understood the business of business. After all, he was privy at the highest levels to the inner workings of one of this country's most important companies, the Pennsylvania Railroad. Having sold bonds on both sides of the Atlantic, he was comfortable speaking to presidents of companies and banks. Occasionally, he was present when Abraham Lincoln visited the telegraph office in Washington that was under his supervision, and he spoke with the president during his railroad service in the Civil War.[115]

Carnegie was well prepared for this first pooling meeting. After hearing of his allotment, he informed the other company representatives present that it was unacceptable. He told them that he was a stockholder in each of their companies and he knew the exact amount of the large salaries and expenses paid to them as presented in their annual reports. He enlightened them with the details of the small salary and no expenses received by the president of the Edgar Thomson Steel Company. Since it was a privately held company, the other members would not be aware of his salary. He shaded the truth, as he often did, when he told them what it cost to make rails at ET. He threatened to undersell all of them, which he was capable of doing, unless he was given as high a percentage of the market as those at the top. (An example that he was capable of doing this can be found by comparing tables 5.3 and 5.5 for 1877.) Records show that his demand was granted. Beyond his role as the principal investor, Carnegie often served as a salesman and acted as the chief promoter for the company, helping in these ways to fulfill the destiny of the Carnegie Steel Company. To the consternation of many, no pool could expect to be successfully arranged without his participation, nor would any existing pool be expected to continue after he left.[116]

Table 5.3. Production Cost per Ton of Rail at Edgar Thomson Works vs. U.S. Selling Price

1877	*Cost at ET Works in USD*	*Price at Mills in USD*
January	44.67	49.00
February	44.89	49.00
March	44.10	49.00
April	43.58	49.00
May	45.63	47.25
June	42.28	46.50
July	44.87	45.25
August	42.55	44.75
September	43.83	44.00
October	42.00	42.25
November	40.13	40.50
December	40.35	40.50

Source: James Howard Bridge, *The Inside History of the Carnegie Steel Company* (New York: The Aldine Book Company, New York, 1903), 97.

Table 5.4. Carnegie Profits vs. Rail Price for Selected Years

Year	*Carnegie Steel Co. Profits in USD*	*Average Price of Rails in USD*
1883	1,019,233	37.75
1884	1,301,180	30.75
1885	1,191,993	28.50
1886	2,925,350	34.50
1887	3,441,887	37.08
1888	1,941,555	29.83

Source: James Howard Bridge, *The Inside History of the Carnegie Steel Company* (New York: The Aldine Book Company, New York, 1903).

Table 5.5. Average Yearly Rail Prices, 1867–1890

Selling Price of Steel Rails per Ton in USD		
Year	*Yearly Average*	*Notes*
1867	166.00	First commercial Bessemer steel made in USA at Steelton.
1868	158.50	
1869	132.25	
1870	106.75	
1871	102.50	
1872	112.00	Lucy and Isabella furnaces begin operation.
1873	120.50	
1874	94.25	
1875	68.75	Edgar Thomson Bessemer begins operation, August.
1876	59.25	
1877	45.50	
1878	42.25	
1879	48.25	
1880	67.50	Edgar Thomson "A" Furnace starts operation.
1881	61.31	Pittsburgh Bessemer Steel Company begins operation, March.
1882	48.50	
1883	37.75	Carnegie buys Pittsburgh Bessemer Steel Company.
1884	30.75	
1885	28.50	
1886	34.50	Homestead, #1 open hearth shop opened, October.
1887	37.08	
1888	29.83	First commercial basic steel made in USA at Homestead.
1889	29.25	
1890	31.75	

Source: James M. Swank, *History of the Manufacture of Iron in All Ages* (Philadelphia: The American Iron and Steel Association, 1892), 514.

Table 5.6. American Bessemer Rail Output, 1867–1876

Bessemer Rail Production	
Year	*Tons*
1867	2,550
1868	7,225
1869	9,650
1870	34,000
1871	38,250
1872	94,070
1873	129,015
1874	114,944
1875	290,863
1876	412,416

Source: J. S. Jeans, *Steel: Its History, Manufacture, Properties and Uses* (London: E. & F. N. Spon, London, 1880), 146.

Table 5.7. Railroad Track Miles in the United States

Year	*Miles Steel Rails*	*Miles Iron Rails*	*Total Miles*	*Percent of Total in Steel*
1880	33,680	81,967	115,647	29.1
1881	48,984	81,471	130,455	37.5
1882	66,611	74,267	140,878	47.3
1883	78,411	70,690	149,101	52.6
1884	90,162	66,252	156,414	57.6
1885	98,013	62,493	160,506	61.0
1886	105,630	62,322	167,952	62.9
1887	125,319	59,586	184,935	67.7
1888	130,388	52,979	191,367	72.3
1889	151,578	50,510	202,088	75.0
1890	167,458	40,694	208,152	80.4
1891	174,775	39,756	214,529	81.5
1892	182,711	38,918	221,629	82.4
1893	190,718	37,185	227,853	83.7
1894	197,491	35,264	232,755	84.9
1895	206,381	28,650	235,031	87.8
1896	210,290	28,440	238,730	88.1
1897	215,658	26,043	241,701	89.2
1898	220,804	24,435	245,239	90.0
1899	229,646	20,717	250,363	91.7

Source: *Poor's Manual of the Railroads of the United States* (New York: H. V. & H. W. Poor, 1900), V.

Based on the information from table 5.7 for 1880 to 1899, which was taken from the *Poor's Manual* for 1900, approximately 20.8 million tons of steel were required for both the construction of new rail lines and for switching from wrought iron to steel rails. This estimate is based on the weight of 352 thirty-foot, sixty-pound profile rails or approximately 106 tons per mile and does not include the considerable additional steel required for joint bars, bolts, nuts, spikes, locomotives, cars, bridges, and more. Eventually trains became longer, faster, and heavier, requiring locomotives that were bigger and stronger. The sixty-pound profile rail no longer had enough strength to bear the effects of the traffic, and new heavier cross-section rails had to be installed. Ninety-seven-pound rail was not unheard of by the 1890s, meaning still more steel was needed, at least on main lines. Many times bridges also had to be replaced, indicating an exponential effect.

In order to speed up operations in the casting pit, Jones invented a device called a horizontal ingot stripper that he patented in 1889.[117] Although its original intent was to only remove ingots stuck in the mould by using the stripper for all of the ingots, the entire ingot and mould were lifted from the pit as a unit by a crane and placed horizontally on a car. The loaded cars were then sent to a horizontally mounted hydraulic cylinder (the stripper) placed at the same height as the ingot. The piston then pushed the ingot from the mould, saving time by not having to strip the mould from the ingot in the pit and then lifting the ingot and setting it on a car. Using the stripper, the pit could be cleared in one motion.

Casting on cars, the next innovation proposed and introduced, was first used in the United States at the Maryland Steel Company and later adopted at Edgar Thomson in 1890 (see figures 5.6 and 5.7). This scheme totally dispensed with pit work beyond the scope of the teeming operation and eliminated the interference with the converter's work. Special railroad cars with empty moulds already set in place were simply pushed into position by a locomotive at the appropriate time. Once they were filled with molten steel, they were pushed out of the way again and sent to the waiting stripper. Totally eliminating the extraneous pit work yielded increased production through using the capabilities of the third converter that primarily acted as a buffer, and the fourth where it existed. With this advancement, the stripper evolved into a vertical instead of the previous horizontal configuration.[118] L. G. Laureau proposed both of these ideas (casting on cars and the ingot stripper) as early as 1876, but, although his concept was good, the equipment design that he contemplated looked somewhat sophisticated and was probably not practical to use.[119] In the steel industry, if it appeared to be complicated, even slightly so, that was usually an indicator of being impractical.

Figure 5.5. The Edgar Thomson Works four-vessel plant of 1890 was equipped with 10-ton vessels. The leftmost and rightmost converters in the view are lying horizontally and the two middle vessels are blowing. Note the hydraulic pit crane and also shallow pit depth. By 1903 the converter capacity had been increased to 15 tons.
From Scientific American, 1903, courtesy of the Hagley Museum and Library.

Figure 5.6. Three vessels of Bessemer Shop at work in the South Chicago Works of the Illinois Steel Company, later the South Works of U.S. Steel. The image shows the two outside converters blowing and the central vessel pouring. This plant layout is unconventional as the vessels are laid to the back to pour the molten steel into a spout, which returns the liquid to the ladle on a hydraulic crane in the front. The large beam across the front of the vessels is for a charging car used to fill the vessel with molten iron for the next blow. Two hydraulic pit cranes dominate the foreground view. Since the cranes are casting on cars (at left in the photo), the photo is post-1890.
Reproduced with permission, Southeast Chicago Historical Society.

Figure 5.7. The hydraulic pit cranes of the South Chicago Works could rotate 180 degrees from the converters to teem the steel into ingot moulds on rail cars, two stations for teeming on cars are at left. One hydraulic crane is poised to teem and the foreground crane needs to be rotated a further 90 degrees. Earlier this function would have been performed in circular pits nearer the furnace. This problem was overcome by adding an extend and retract cylinder to the ladle arm of the crane.

Reproduced with permission, Southeast Chicago Historical Society.

The demand for iron imposed on Lucy by ET forced the construction of a second blast furnace there. Rapidly, the massive increases in output attained by its converters far exceeded the ability of Lucy to supply its needs even with two furnaces. A decision to build more blast furnaces pointed logically to Braddock, not Lawrenceville. At the time the new facility was constructed in Braddock, the Lucy Furnace Company and the Edgar Thomson Steel Company, although related, were actually independent entities. This would be a source of irritation in the future because some of the owners were not involved in both enterprises and conflicts arose when members of one group profited at the expense of the other.

If it was at Lucy and Isabella where the idea of hard driving was initiated, it was at Edgar Thomson where it was honed to a fine art. The first furnace at ET was designated as "A" Furnace. Based on information from various sources, building it was a mistake from the beginning. Blast Furnace "A" was an old charcoal furnace brought in from Escanaba, Michigan, erected at Braddock in late 1879 and put in operation using coke the following January. At sixty-nine feet in height by thirteen feet at the bosh, "A" was very small when compared to the Lucy stacks (75' × 20') and other high-production American coke furnaces. The furnace came as part of a settlement, when an iron ore mine partially owned by Kloman failed and was purchased for $16,000, dismantled, brought to Pittsburgh, and reassembled under the able leadership of Julian Kennedy, who was brought in to supervise.[120] Kennedy had been the superintendent of the Brier Hill and Struthers Furnaces near Youngstown, Ohio. Just prior to that, he spent a brief stint as a physics instructor at Yale University.[121] Since it was originally planned to produce spiegel or ferromanganese, a large furnace wasn't required to produce the volume of material needed for recarburization of steel when considering the relationship of required spiegel to the amount of iron demanded by the converting department at ET. This was probably where some of the confusion about its dubious construction arose. After making a short run on iron in early 1880, June of that same year saw it relegated to making the spiegel as first planned. In that interim period before spiegel, remarkable things were being done with the "A."

Apparently there was an internal squabble at the steel company concerning the type of stoves and blowing engines to be used at Edgar Thomson. According to Shinn, some of the partners at Lucy felt that theirs was the best technology available—Whitwell cast iron stoves and *simpled* (not compound) blowing engines. Their attitude seemed to be "How could one be expected to better a world record holder?" A committee of three was appointed to visit all but one of the American furnaces using firebrick stoves with the result that the Siemens-Cowper-Cochrane firebrick stove design (usually referred to simply as Cowper), rather than the cast iron Whitwell stoves, were used. Along with that decision, powerful engines of the most modern design were built.[123] Kennedy soon identified some flaws in the internal arrangement of the stoves, and he quickly modified them to improve their heat distribution, reduce pluggage, and correct flow problems.[124]

In an era of extremely abundant and inexpensive fuel in the Pittsburgh region and the consequential immense waste that typically accompanied that fact, Kennedy made a somewhat profound statement for the time (1880), which I am sure was lost on most of his audience. He said "that heat was money and should not be wasted unnecessarily even around a blast furnace."[125] The economic and wise use of raw materials (for then) was a prevailing theme of the Carnegie Company and is a credit to the men who managed their operations. At a time when coal could be bought in quantity for cents per ton, efficient use of fuel was often not a great concern because the net savings resulting from the big effort to reduce fuel usage was not large. There was also a lack of understanding about combustion, demonstrating the belief that if some fuel is good, then more is better. Often this perception culminated in those obscene views of stacks with their heavy outpouring of opaque, dense, black smoke, which was an otherwise common-enough sight for more understandable reasons. Most men could not comprehend that when too much fuel has been added, a point is reached where it actually begins to *take away* heat, rather than add it. Typically, dense smoke spewing from every stack was perceived to mean that men were at work; work meant prosperity, therefore smoke was good. Kennedy was surrounded by many intelligent men, who had minimal formal education or were self-educated. Having been trained in physics at a prominent university, he had a unique perspective.

"A" Furnace had only about 40 percent of the cubical capacity (internal volume) of Lucy 2 (see comparison in table 5.9). Still, Kennedy was able to take advantage of the more powerful blowing engines designed for the new "B" Furnace, which weren't being used because "B" was still under construction at that time. Borrowing the larger blowing machinery from "B," together with adopting the use of the new larger, more-efficient firebrick stoves constructed for "A," permitted "A" to be vastly *overblown*—that is, hard driven with fifteen thousand cubic feet of air per minute (cfm). This was almost as much air used at Lucy, which had more than twice the furnace volume. The results were phenomenal. In five months' time more than 10,400 tons of iron was produced, despite frequent problems with hanging on this tiny furnace. In the last month of operation, before switching to spiegel, only 1,945 pounds of coke were used for each of the 2,226 tons of iron made, equating to seventy-one tons output per day over the five-month run on iron. Lucy 2 averaged about ninety-one tons a day for all of 1878, but it was twice as big and a world record holder as well. It would be another five years before any large furnace beat "A's" record on fuel consumption.[126]

There was a conscientious effort by the Carnegie Company to provide adequate machinery to complement the production of iron and steel at Edgar Thomson. Shinn visited many plants, where he was invariably told by one manager or another, "we are a little short on steam this morning" . . . or, "we are short of blast to-day" . . . or, ""we we have been a little short of hot-blast to-day" . . . or, "we are very short of gas for our boilers." Shinn was determined that the furnaces at ET were never short of any of these "things."[127] Therein lies an important reason for the successful wedding of the blast furnaces and converters at ET: a determination to be the best. The plant was known for being a well-equipped facility.

At 80' × 20', "B" Furnace was slightly larger than Lucy in both height and volume. It was put on blast in April 1880, a bit before "A" was changed over from smelting iron to making spiegel. In an attempt to produce a better material profile in the shaft, this furnace initially had an unusual double (annular ring) bell for loading it that was later replaced by a standard-shaped bell because it had no apparent effect. The hearth was eleven feet in diameter. In America no furnace had ever been built with a hearth this large. According to Shinn, the increased size was deduced through experience, when it was noticed that smaller hearths normally eroded away to a much larger size by the end of a furnace campaign.[128] Also, the experience of the Catasauqua Furnace (Crane Iron Company) where they had gradually enlarged the hearth to this dimension was known. Firebrick stoves of Cowper's design—large, well-built blowing engines and a more than adequate supply of boilers—were also included. With an ample supply of ancillary equipment, this furnace was hard driven similar to the practice developed on "A," but with 30,000 cfm of air, or slightly less than twice its cubical capacity per minute, which was more air than had ever been used on a furnace before. The ratio of air to cubical capacity used on "A" had been slightly greater than two times. The output of "B" reported for that month was 2,723 tons. Six months later it produced 4,722 tons or almost double the amount of April. This was more iron than had ever been made anywhere in the world. In fact, such output was so unusually high that many either didn't believe it or thought that the numbers were being biased by charging and melting large quantities of scrap in addition to the iron ore, similar to the operation of a cupola.[129] The average monthly production for Lucy 2 in 1880 was only 2,828 tons. "These furnaces [ET] at once leaped to the front as pig-iron producers, and have maintained that position—with but one brief interruption—ever since," wrote James Gayley.[130] The mention of hard driving was first broached in literature covering Isabella and Lucy during the 1870s. In reality this concept found its basis and definition from the work done by the short-lived operation of the diminutive "A" Furnace, while working on iron, and was then carried forward by the startup of "B." From 1879 until 1890 there were nine furnaces built at ET in addition to the redesign and rehabilitation that occurred about every three years when linings wore out. Two more furnaces were added in 1902 to 1903 for a total of eleven, all of which were assigned names from "A" through "K."

> Modern iron making in America began when, in 1881, the long-doubted [rumor] became certainty, that the late Captain William R. Jones and Julian Kennedy, had, by means of high heats and large volume of blast succeeded in more than doubling the output of the Edgar-Thomson Furnaces without altering the plant . . . It is along the lines laid out by these men, and on the foundations largely built by them, that the iron industry of the United States to-day is developing.[131]

It may appear that the success of hard driving was merely due to blowing a large volume of air (*wind* in today's terminology) into the furnace. In actuality, creating these bigger outputs represented a far more inconspicuous challenge than that conveyed by the idea of building bigger blowing equipment. As previously stated, logic doesn't always work in the way that it seems that it ought. Also, simply throwing money at a project does not usually solve it, particularly when understanding is lacking. With the exception of "A," the dimensions of each of these furnaces were based on what was thought to be successful from a past design, or through analysis of and extrapolation from previously acquired operational information. Furnace dimensions were changed (both increased and decreased) at the hearth, bell, bosh, and stock line. The location of the bosh (its height in the stack) was also adjusted; this altered the angle of the stack to a greater degree than simply manipulating the furnace dimensions. This affected the flow of material. The volume and pressure of air blown, as well as the temperature of the blast, was increased or decreased to determine its effect. Stoves were also modified. Sometimes changing the furnace blowing parameters tested the capacity limits of the engines to blow and/or the stoves to heat the air. The number and location of the tuyeres was changed. Cooling plates were added to the bosh and hearth walls; their numbers, locations, type, design, and method of installation were changed to determine how these alterations affected production. The furnace linings were inspected when they were blown out to ascertain the effects of hard driving from the wear patterns in the brick or from carbon deposition on them. So the changes being made were not arbitrary, though they were often just based on estimates from the best knowledge of the time.[132]

The men at Edgar Thomson were driven by their furnaces just as they drove them, but theirs was a burning desire to learn, to improve, and to succeed. Kennedy provided some inkling into the intricacies and pitfalls of making even subtle changes when he alluded to "the great effect produced by very small, or apparently very small, changes in the lines [shape] of the blast furnace" because these modifications were responsible in part for the increased output, and also that "our fast driving was not due altogether to our desire to drive fast. We often tried a little slower rate of driving, but then found that the furnace would not work, and so we were obliged to drive as we did. I ascribe a great deal of this to the difference in the lines of the furnace then from what they are now. These differences do not seem very great, but still they made a marked difference in the result."[133]

> In the period covered by the last decade there are three steps in the development of American blast-furnace practice that might be mentioned—first, in 1880, the introduction of rapid driving, with its large outputs and high fuel-consumption; second, in 1885, the production of an equally large amount of iron with a low

fuel-consumption, by slow driving; and third, in 1890, the production of nearly double that quantity of iron, on low fuel-consumption, through rapid driving —James Gayley, Braddock, PA.[134]

The data provided in table 5.9 and displayed in the accompanying graph (chart 5.1), which was derived from a portion of that data, portray the dramatic increases in iron output attained by the men at Edgar Thomson by the end of 1890 (see figure 5.8). E. C. Potter was a highly respected former general manager of the South Chicago Plant of the Illinois Steel Company (South Works of the North Chicago Rolling Mill Company) and the closest competitor of the ET blast furnace in this era. He was the person responsible for briefly deposing ET from the position of world record holder. Potter's disposition on the matter was that "dating from the blowing in of the Edgar Thomson 'A' furnace in 1879 and continuing down to the present day [1893] . . . marked the beginning of a new era in the science of iron-smelting, if, indeed, it was not the very birth of science as applied to this industry in the United States."[135]

Whenever there were steelworker strikes at Edgar Thomson, company-imposed production stops (lockouts), economic conditions that caused

Chart 5.1. Iron Production. Shows the dramatic rise in iron output per day for selected furnaces in Pittsburgh (all are Carnegie Co. except Isabella) and an associated precipitous drop in fuel consumption that accompanied that rise. Although there appears to be brief dips and stalls by year in the overall trend, it in fact displays increasing production and decreasing coke consumption when comparing, the same furnace to its future rebuild.

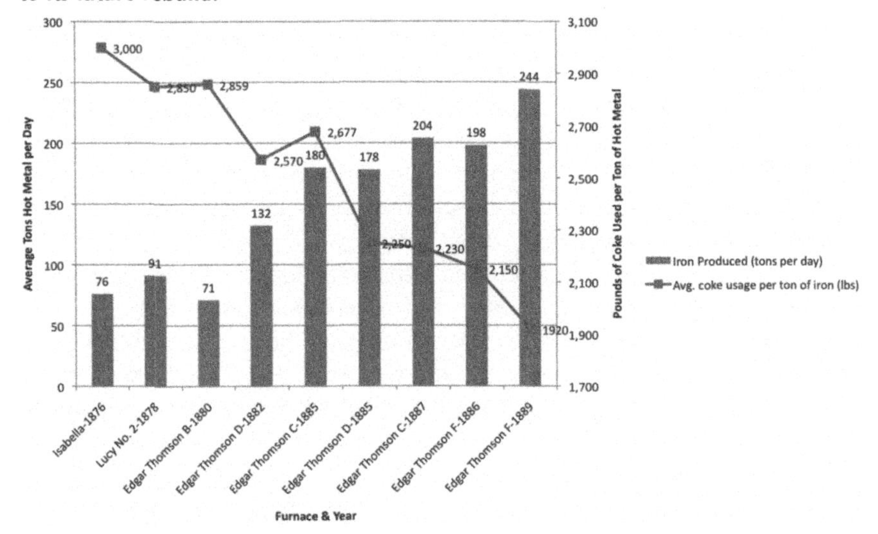

Data from Table 5-9, James Gayley, "The Development of American Blast Furnaces With Special Reference to Large Yields," *Transactions of the American Institute of Mining Engineers*, XIX, New York, 1891, p. 956.

Figure 5.8. A photo overview of the Edgar Thomson blast furnace plant in 1886, during the Carnegie Steel Company's second period of blast furnace transformation (Lucy being the first) according to Gayley's description above. You can note that the blast furnaces are flat topped and hand fed by elevators.
Thomas & Katherine Detre Library and Archives, Sen. John Heinz History Center, Pittsburgh.

shutdowns, coke shortages, coal miner and coke worker strikes, or weather conditions, such as floods that affected the availability of raw materials, all negatively impacted the coke rate on each individual blast furnace. From that point forward in the campaign, coke consumption on that particular furnace or furnaces was always higher, until it was blown out and relined. Not once in the decade or so run of the furnaces under discussion here did they operate even one furnace without stopping and/or banking. So the true maximum potential of their operation was never fully known. Peak output usually occurred during the first twelve months of operation, and these are the figures most often given for the annual output.[136]

When making comparisons between furnaces at different locations or often even within the same plant, when trying to make all things equal, the facts show that they just never are. Since no one used the exact same blends of ores, the furnaces would act slightly different, the lines of the furnaces were never quite exactly the same, and as Kennedy alluded, even apparently minor changes had major effects.

Sometimes, the cost of raw materials ruled the decision-making process. Potter noted that as Edgar Thomson had immediate access to coke at low prices, there was an advantage in this area as far as fuel rates were concerned.

The South Chicago plant that he led had to pay large freight costs on coke, even though its supply, also from Connellsville, was priced competitively. So coke usage was monitored very closely because the impact of the cost of transportation was huge. On the other hand, both Chicago and Pittsburgh received most if not all of their ore supplies from the Lake Superior region. South Chicago could use the ore directly as it was unloaded from the boat, while ET had to reload it onto trains to ship it to Pittsburgh. Other plants like Cambria, Harrisburg, and Bethlehem were at a bigger disadvantage because they had to ship both of these commodities some distance by rail. British and European ores were frequently composed of material containing 50 percent iron and lower as compared to Lake Superior's 60 to 65 percent or better. Continental cokes were made in "slot type" or "Belgian" ovens and exhibited differing qualities. So the differences were most often small but many. This made close comparisons difficult, if not impossible.

> The largest furnaces of this period [1874] were 75 feet high and 20 feet in diameter at the boshes. About 20 years later, towards the close of this period, the largest furnace in the United States, at Braddock, was 90 feet in height and 22 feet in diameter at the bosh. The growth in capacity, therefore, was largely due to improvements in furnace practice rather than increased size.[137]

This hypothesis is supported by a response made at Gayley's presentation on American furnace development at a meeting of the American Institute of Mining Engineers. The subject was posed somewhat differently by G. J. Snelus, a respected British metallurgist and plant manager, who said, "I have always been an advocate of rapid driving, but my difficulty has been that I could not drive with any real rapidity. The men would not be driven, the manager would not be driven, and the furnace would not be driven, because we did not have the plant and could not get it."[138] Period British furnaces were of similar sizes or even larger than those at ET. His comment reveals the positive attitude and work ethic of American workers and plant managers (what we would call today "can do"), at least as far as the Edgar Thomson works was concerned.

One further comment will be made here about the "A" Furnace and its run on spiegel. There was only one other manufacturer in America, so the availability of spiegel in the United States was quite limited. Julian Kennedy convinced Carnegie that the company's ore supply was adequate for smelting ferromanganese instead. No other plant in the United States made it, and before "A" Furnace began production, the entire country was reliant on foreign sources. For many years Carnegie became the sole supplier in America and eventually corralled the entire market, but despite his monopoly the price dropped from $80 to $50 per ton for this important commodity.[139]

Table 5.8. Pittsburgh (Allegheny County) Blast Furnaces to 1890

Year Built	*Name / Designation*	*Owned By*	*Height Ft.*	*Bosh Dia. Ft.*	*Capacity Tons per Year*
1859	Clinton	Graff Bennett & Co.	45	12	15,000
1861	Eliza No. 1	Laughlins & Co. (J&L)	60	17	. . .
1861	Eliza No. 2	Laughlins & Co. (J&L)	60	14	30,000§
1863	Superior No. 1[a]	Superior Iron Co.	45	12	. . .
1868	Superior No. 2[a]	Superior Iron Co.	45	12	25,000§
1865	Shoenberger No.1	Shoenberger, Blair & Co.	62	13	. . .
1865	Shoenberger No. 2	Shoenberger, Blair & Co.	62	13	48,000§
1872	Isabella No. 1	Isabella Furnace Co.	75	18	. . .
1872	Isabella No. 2	Isabella Furnace Co.	75	20	80,000§
1872	Soho	Moorhead, McClean & Co.	65	19	31,000
1872	Lucy No. 1	Lucy Furnace Co. (CSC)	75	20	. . .
1877	Lucy No. 2	Lucy Furnace Co. (CSC)	75	20	85,000§
1879*	Furnace A	Edgar Thomson Steel Co. (CSC)	65	13	25,000
1879*	Furnace B	Edgar Thomson Steel Co. (CSC)	80	20	45,000
1879*	Furnace C	Edgar Thomson Steel Co. (CSC)	80	20	45,000
1882	Furnace D	Edgar Thomson Steel Co. (CSC)	85	20	
1882	Furnace E	Edgar Thomson Steel Co. (CSC)	85	20	
1883	Carrie	Pittsburgh Furnace Co.	70	18	40,000
1886	Eliza No. 3	Laughlins & Co. (J&L)	60	17	30,000
1886	Furnace F	Edgar Thomson Steel Co. (CSC)	90	22	
1887	Furnace G	Edgar Thomson Steel Co. (CSC)	90	22	
1890	Furnace H	Edgar Thomson Steel Co. (CSC)	90	22	
1890	Furnace I	Edgar Thomson Steel Co. (CSC)	90	22	

*Started operation 1880

§ Total output for both furnaces.

[a]Later renamed Edith and replaced with a single stack 70′ × 16′ built in 1882 and listed in 1888 as having a yearly capacity of thirty-five thousand tons per year. Later owned by the National Tube Company after the Superior Company met financial difficulties.

Source: "Annual Report of the Secretary of Internal Affairs of the Commonwealth of Pennsylvania, Part III," *Industrial Statistics* XXII, no. 10 (1894), D. 98; William F. Switzler, *Commerce of the Mississippi and Ohio Rivers*, Report of the Internal Commerce of the United States, Treasury Department, Government Printing Office, Washington, 1888, 402; *Transactions of the American Institute of Mining Engineers* VIII, Easton, Pennsylvania, 1880, 370. (This table was assembled by combining data from tables contained in the three documents cited. There is some conflict as the later citations reflect the more modern state of the facilities—size, output, and more—but the original construction size and date were retained.)

> The fundamental principle which has been carried out at Pittsburgh [CSC] was to destroy anything from a steam engine to a steel works whenever a better piece of apparatus was to be had, no matter whether the engine or the works was new or old, and the definition of this word "better" was almost entirely confined to the ability to get out a greater product and get it out uninterruptedly. Such a course involved the expenditures of enormous sums of money, it involved the constant return of profits into the business, it involved many mistakes, but it produced results, and the economics arising from the increased output soon paid for the expenditure. This example has exerted a great effect upon other American steel works, and is also felt to a considerable extent abroad.[140]

In 1890 the United States surpassed Great Britain and became the largest iron producer in the world.[141]

We can extract an appreciation for the enormous strides that were made in the fifty-year span of ironmaking in this country, starting from the 1840s era, by comparing American cold-blast charcoal furnace practices to that of the state-of-the-art Edgar Thomson furnaces as they were in November of 1894. The total number of employees associated with making iron at ET for that entire month was 1,850 men, during which time roughly eighty-six thousand tons of iron were smelted. It was estimated that 1,500 coal miners (who would have had to have dug approximately 132,000 tons of coal based on 0.65 ton of coke from one ton of coal) and one thousand coke workers were required to provide sufficient coke (around eighty-six thousand tons were required at a rate of one ton of coke to one ton of iron output, but actually it was somewhat less with a ratio closer to about 0.9). A comparable estimate of the number of charcoal furnaces needed (circa 1840) to create the same output would have been somewhere between eight hundred and one thousand furnaces, which would have required twelve thousand to eighteen thousand men to operate. By comparison, there were only eight furnaces making iron at ET in 1894, the ninth, "A," was making spiegel. They would also have needed about seven thousand men cutting trees (at a rate of about one square mile per day or thirty square miles of forest for just one month!) and nine thousand colliers to make the charcoal, together with one thousand draft animals to haul the wood. This equates to a total of 4,300 men at Edgar Thomson making coke iron versus twenty-eight thousand to thirty-four thousand men needed for manufacturing the same amount of charcoal iron each and every month, or roughly 600 to 800 percent more men than were required for coke iron. This does not include the number of men to transport coke from the mines to the mill by train, although this was not thought to be significant.[142]

The innovations spilling forth from ET had not yet been concluded.

Table 5.9. Improvements Made to Iron Production at Selected Pittsburgh Furnaces, 1876–1890

					First 12 Months of Output			
Furnace Designation	*Initial Year of Operation for Campaign*	*Cubical Capacity of Furnace Cubic Feet (ft^3)*	*Vol. of Air Blown (Wind) ft^3 per Minute (cfm)*	*Total Tons Furnace Output during Campaign*	*Total Iron Produced (tons)*	*Avg. Iron per Day (tons)*	*Avg. Coke Usage per ton of iron (pounds)*	*Furnace Volume per Ton of Iron Made (ft^3)*
Isabella*	1876	15,000	Unknown	117,575	28,000	76	3,000	197
Lucy No. 2	1878	15,400	16,000	92,128	33,552	91	2,850	169
Edgar Thomson A	1880	6,396	15,000	. . .	. . .	71[a]	2,400	90
Edgar Thomson B	1880	17,868	30,000	112,060	48,179	132	2,859	135
Edgar Thomson D	1882	21,478	27,000	90,317	65,947	180	2,570	119
Edgar Thomson C*	1885	16,680	31,000[b]	118,000	64,998	178	2,677	90
Edgar Thomson D*	1885	18,950	22,000	150,374	74,475	204	2,250	92
Edgar Thomson C*	1887	17,230	24,000	203,050	72,554[c]	198	2,230	87
Edgar Thomson F	1886	19,800	27,000	224,795	88,940	244	2,150	81
Edgar Thomson F*[e]	1889	18,200	25,000	. . .	113,000[d]	310	1,920	59

*Subsequent campaigns on a particular furnace reflecting the effects of physical and/or operational practice changes after rebuilding/relining.

a—Estimated.
b—After running nine months, the volume of air was reduced to 28,000 cfm.
c—The second twelve full months showed an output of 83,219 tons using an average of 2,396 pounds of coke per ton.
d—Estimated from record made to date (see reference e below).
e—Actual production for the first twelve months of blast of Furnace "F" for the campaign beginning in 1889 showed an output of 113,526 tons of iron (311 tons/day) with an average coke consumption of 1892 pounds per ton.

Source: Adapted from James Gayley, "The Development of American Blast Furnaces with Special Reference to Large Yields," *Transactions of the American Institute of Mining Engineers* XIX, New York, 1891, 956.

Using a cupola was an expensive, albeit necessary, proposition for a Bessemer steelmaking operation, and men were always pursuing schemes to eliminate them, or at least limit their impact on costs that were considerable. First there was the expense incurred at the blast furnace for all of the pig bed men required to prepare the pig beds; to cast, cool, and break them; and finally to grade and load the pig to be transported to the steel plant. Second, there was the expense of the cupola and its operation, the men to load and operate the vessel, and, more especially, for the quite significant fuel consumed to melt the pig.

> [The] H. C. Frick Coke Company introduced the crushing of coke with a view to taking the place of anthracite coal [in foundry practice at least]. . . . It has been found that a given weight will melt 30 to 50 per cent more iron in 20 per cent less time than the same weight of anthracite. The coke, of course, requires less blast than other fuel, and melts iron hot and soft, more scrap can be used. The reduced quantity of slag to be dealt with, and the absence of clinker, are other advantages, while coke is easy on the cupola linings. It has been computed that while 1 lb. of coal will melt 9 to 14 lbs., and that while a cupola of economical size will pour off the iron melted with coke in two hours, it will take three hours when using coal.[143]

According to Howe, it took more than a tenth of a ton of coke (two hundred pounds) per ton of iron melted, but at good operations could be as low as one hundred pounds. He attributes the lower requirement at faster operating plants to the consumption of silicon that was incorporated within the matrix of the iron. The silicon was burned in the cupola, which added heat to the process.[144] So the fuel expenditure of the cupola was the equivalent of adding a 5 to 10 percent penalty to the coke rate at the blast furnace that they were so ardently trying to trim at ET. Jones pursued a different way to reduce the impact of the cupola. At the onset of their use, these *shaft furnaces*, as they were sometimes called, were emptied (dumped) after a given number of tons were produced, taking only a few hours.[145] Depending on the practices at a particular location, this was modified to about every twenty-four or forty-eight hours of operation, while producing sixteen tons per hour on average. This was not based on production, but rather hours of use.[146] Jones developed methods by which the furnace was able to run 141 continuous hours without dumping.[147] The act of dumping was simple. Simply pull a single pin or knock down a supporting post or posts underneath that held closed a pair of semicircular, overlapping, hinged cast iron doors, the diameter of the cupola that when closed formed the floor of the furnace. When the support was removed, the doors swung open due to gravity, and everything inside the shaft dropped right out of the bottom. To avoid dumping, if you could possibly do so, was significant. There were some fairly large repercussions each time a cupola was dumped. For instance, a large mass of flaming, red-hot coke, ash, molten iron, and slag was deposited on the floor of the cupola house, directly below the furnace, together with all of the smoke and noxious fumes released each time it was emptied. These tons of mess had to be extinguished, hand shoveled, and wheeled out of the way by laborers before the cupola could be prepared for the next campaign. In some diagrams of Bessemer houses that had highly elevated cupolas to feed the converters, the floors beneath the cupolas were sloped. This diverted the mess away from the furnace to the outside of the house so that the cleanup need not be done immediately, nor did the smoke and fumes impact the operations inside of the building. The shaft had to be inspected and patched if necessary, and the doors closed, pinned, or propped shut. These were hellishly heavy doors that had to be manhandled. The floor of the furnace was made by evenly spreading a thick layer of sand over the closed doors to form the base of the shaft. Then a large pile of scrap lumber or kindling was loaded inside and set afire, and coke was added for ignition. (Coke is a strange animal. It is often very difficult to ignite, but once it starts to burn, it seems nearly impossible to extinguish.) More coke was then added, followed by pig metal, and the operation routine was restarted. The aforementioned 141 hours of continuous running time attained by Jones was significant, because it closely matched the typical number of operating hours (eleven to eleven-and-one-half turns of twelve hours duration) for the Bessemer converters at Edgar Thomson for an entire week.[148] The remaining twenty-seven or so hours (about thirty to thirty-six hours for the converters) in the week consisted of down time that was usually committed to the maintenance of the cupola and converter vessels to prepare them for operation during the next week's work.

The principal benefit of not using the cupola method, if it were at all possible, was quite obvious. Using molten iron directly from the blast furnace saves all of the fuel required to remelt it.

The direct metal process, or simply the direct process as it was known, was developed in Europe around 1871 and came into use in England in 1877. Plants that were thought to have excellent iron smelting operations—ones that produced a consistent quality product—converted to the direct method. In the United States, the South Chicago Works was the first to use it in 1882, followed by Edgar Thomson later that same year.[149] The key words used above are "thought to have," but in point of fact no one was completely capable. In 1890 Howe stated that various managers estimated the savings from using the direct method to be between $0.50 and $1.00 per ton.[150] This is 2 to 5 percent of the cost of rail sold in the $20 to $30 per ton range. This was fairly significant, but it was fraught with many problems. Large variations in the amount of silicon resulted in:

- inaccurate blowing due to inconsistently estimating endpoints on the blow
- blowing too long due to high silicon, thus reducing productivity
- hot heats, requiring the addition of scrap and thus reducing productivity
- cold heats from too little silicon leading to *skulling* in the vessels and ladles (Inches thick, a skull was a solidified mass of steel forming in and/or adhering to the brick lining of a ladle, but it could also occur on the metal cases of other vessels, like converters. Once formed, it often had to be removed before the vessel could be used again. When removed from a ladle, some thought that it resembled a large human skull, as it was usually upside down, and it often weighed many tons. Sometimes, it was the entire heat.) This increased costs.

For an assortment of reasons, this variability in the iron resulted in:

- increased production of second-quality rails
- more scrap heats
- higher converter losses due to longer blowing
- more butt end ingots
- higher consumption of manganese

These were only some of the negative aspects encountered.[151]

In 1889 Jones was granted a patent for a device that he termed a mixer. It consisted of a large refractory-brick-lined, metal-plate vessel, originally designed in the shape of a wedge or a lopsided trapezoidal box. Later, though, it was usually in the form of a cylinder. The prototype vessel was capable of holding about 80 tons of molten iron and quickly was enlarged to hold 150 tons or more. The mixer was devised as an intermediate vessel that was interposed between the blast furnace and the Bessemer converter. Molten iron was brought from the blast furnace in groups of fifteen-ton ladles that were spotted, then tilted (tipped forward) and poured into the mixer through an opening at the top level. As iron was needed for the converter, it was extracted from the mixer, which was mounted on a pivot, by tilting it forward with mechanical or hydraulic machinery to pour the molten iron out through a nozzle into a ladle waiting on the lower level. Depending on the size of the converter, six- to twelve-ton lots were dispensed each time to produce steel (see figure 5.9). At first this idea does not seem to be anything extraordinary, but a single instruction given to the man or boy operating the vessel made it revolutionary. According to testimony given at a patent infringement hearing, "A chalk mark . . . [was] on the side of the mixer . . . [and when pivoting the vessel] they didn't allow them to run that chalk mark below the floor. When it came to the floor, they still had retained in the mixer about 175 tons."[152] That simple

Figure 5.9. The Jones Mixer at the Edgar Thomson Works. Newly tapped iron from the blast furnace is being poured into the mixer trough an opening in the top of the vessel. An empty ladle can be seen on the lower level at the left waiting to receive a charge of iron to take to the Bessemer Shop to be converted into steel. The mixer discharged the iron by tipping forward and pivoting about the bearings seen to the right of the post in the foreground. The container was never emptied so that the variability in the chemistry of the iron from a new cast could be blended out in the reservoir of remaining material. Engineering, London, Vol. 24, 1903, p. 406.

instruction, not to let the chalk mark on the mixer go below the floor, helped to revolutionize the steel industry. To avoid emptying the mixer vessel, which could happen if the chalk mark was allowed to go below the floor, guaranteed that there was a large enough reservoir of iron retained within the chamber with which new batches of iron from the blast furnace would mix to average out the high variability in the chemistry of the new iron (most importantly the silicon and sulfur). The variations in the quality of the iron were readily apparent not only between the outputs from different furnaces but also within the same cast of a single furnace, and sometimes even within the same ladle.

Reduction in the variability of the chemistry of the iron was the basis for the patent. When Jones put his mixer into operation, it was unique. Less than a decade later, if a steel company with blast furnaces did not have a mixer in their plant, it was in the extreme minority. This was a simple yet elegant solution to the problems of using iron directly from the blast furnace and was also used for replacing cupola metal in the converter. Another benefit was that it also smoothed out peaks and valleys in production. Little cost yielded large rewards. This was certainly a good idea, but was it revolutionary?

An obscure yet monumental consequence of the invention of the mixer was that it forced a change in the concept of what an efficient steel mill had to be and what it looked like. A matter of the proverbial not seeing the forest for the trees, it's not an obvious change, because it already existed in the same form in which it must continue in order to function successfully and reap the benefits under a new concept. Little appeared to be any different or special, but eliminating the variations in the iron had an enormous impact in a plant the size of Edgar Thomson, when compared to use of cupola metal. Some of these items were:

- the thousands of man-hours per month saved through the elimination of all of the laborers required for the pig beds and cupolas (At the time of its invention, there would have been eight pig beds at ET, plus the uncounted one that was used for casting spiegel on "A," and from four to six cupolas for the converters.[153])
- coke savings of 5 to 10 percent based on the fuel rate at the blast furnace
- improved quality due to the reduced sulfur load resulting from not contacting coke in a cupola, and other benefits
- the reduction of variations in the chemistry of the iron resulted in more consistent operation of the converters, which increased productivity

All of these benefits were tangible, measurable, and recordable outcomes. When Edgar Thomson began steelmaking operations, it was a stand-alone entity that relied upon cupolas for hot metal that was shipped in as cold pig from Lucy Furnace, some nine miles distant as the crow flew. When more than two blast furnaces were needed to supply the iron for ET, it made more economic sense to build the additional capacity in Braddock to save shipping expense, even though they continued to use cupola metal. In fact, for a short time the Lucy Furnace Company was sold to Wilson, Walker and Company.[154] When the use of the direct process for using iron as it came directly from the furnace was instituted, fuel was saved but problems arose with the fluctuations in the chemistry of the iron, increased ingot scrap, as well as other associated costs, like:

- increased maintenance on ladles
- more skull scrap
- more secondary product, and more, associated with running converters on direct metal

When the Jones mixer was used, in addition to savings accrued by using direct metal, these resulted in added economic benefits from eliminating detrimental aspects incurred from using the direct process, while simultaneously increasing productivity and reducing the labor force.

If the blast furnaces and the steel plant were in separate locations, the benefits of the mixer were lost. In order to gain them, a modern high-production steel plant at the forefront of the industry had to consist of a blast furnace in conjunction with a Bessemer converter plant adjacent to it, or nearly so, with a mixer located in between. That was the revolution! Edgar Thomson had been operating on direct metal for some years before the mixer was adopted, so it did not appear to be much different than it had been. Yet the revisions to procedures introduced by the mixer totally changed the way in which things had to be done. Similar to the invention of the phone, the car, and the computer, they are not necessarily needed, but once experienced, the benefits demand their use. All of these everyday devices were once revolutionary.

Almost immediately, the Cambria Iron (Steel) Company began using the mixer in infringing upon the patent, so much so that they even used a chalk mark to denote where the maximum lower limit of travel for their mixer was permitted.[155] A number of other steel companies began infringing as well. Cambria argued that the mixer was simply a holding vessel that were proliferate in many industries, including foundries, and were used as surge vessels to dispense the molten product as it was needed (for each individual casting in a foundry, for instance). Carnegie sued, and one case (*Carnegie Steel Co. v. Cambria Iron Co.*) was heard before the U.S. Supreme Court that ruled in Carnegie's favor in 1902.[156] The Court ruled that the Jones mixer met the criteria that: it was of large size (one hundred tons or more); was a covered vessel to prevent the entrance of cold air; that it had a stop so that it was not tilted too far and permitted to empty its contents; and that a large enough quantity of metal was retained to reduce the variability of product received from the blast furnace that was then discharged for the converters to consume. No other patents or processes cited met all of these criteria.[157] However, by then Jones was dead. The patent had passed to the Carnegie Steel Company, sold to them by his widow, and was then transferred from CSC to United States Steel before the 1902 ruling came. United States Steel was formed through the merger of Carnegie with others after CSC was sold to J. P. Morgan in 1901. The estimated value of the patent infringement was in the millions of dollars, but the actual value is not known, because some of the violating companies became part of United States Steel in the merger.

Bill Jones did not live long enough to learn of the massive impact that his mixer had on the world of steelmaking. Even considering the many advancements that have been made leading up to the present state of the industry, mixers continue to be used well over a century later, including at Edgar Thomson. The quotations cited from Campbell, Potter, Howe, Kennedy, and Gayley in this chapter are direct references to Jones's policies and concepts that evolved from them through his encouragement, just as Jones's skill grew through his association with the innovative techniques developed by the Crane Iron Works, Alexander Holley, and George and John Fritz. By the end of Jones's life, the Edgar Thomson Works was widely recognized as the most productive steel plant anywhere.[158] His demise was accompanied by all of the drama and exuberance of his life. Bill Jones died on the firing line, while directing the daily battle to defeat oxygen, carbon, phosphorus, and sulfur in search of steel.

Jones's death on September 28, 1889, at age fifty-nine resulted from an industrial accident at the Edgar Thomson Works. The "C" Blast Furnace had developed problems, as these temperamental beasts were often wont to do. In what some described as an explosion and others recount as a collapse, Jones was struck by material being expelled from the furnace as he watched the men work. The force of the material being ejected knocked him from the cast house, and he landed on the floor of the pit below, where he struck his head. James Gayley stated that he was near Jones when the furnace wall suddenly collapsed.[159] Many times a wall failure was precipitated by the crumpling of a "hang" from above, and the impact of the tremendous weight of all of that raw material crashing down into the hearth would sometimes cause the furnace to burst at its seams. *Railway World* stated that the accident "was caused by a sudden outpouring of a mass of molten metal which succeeded prolonged attempts to drill a hole into the furnace for the purpose of overcoming a difficulty caused by the tap and cinder holes being clogged up."[160] Just two years before in May of 1887, five men were killed on Furnace "E" when the red-hot mass of coke, ore, and stone suspended above broke loose and fell onto them.[161] Jones died two days later from head injuries and burns sustained in the incident, having never regained consciousness. Another man, Andrew Hanneyloi, whom most accounts of the event usually identify only as a Hungarian, standing near Jones was also killed. His fate was far more grisly. The newspaper reported that he was almost totally cremated. A third man from Braddock, named Michael Quinn, also died in the incident.

At a meeting held at the Opera House in town, a crowd of workers estimated to be about two thousand men assembled to decide if they should or should not work until after Jones's funeral. The more honorable path was to work. Braddock's school announced that it would be closed for the day of the service, but his widow encouraged them to remain open.[162] The lead line of a story about the accident in the *Pittsburgh Press* stated: "Capt. William R. Jones, the greatest mill boss in the United States, perhaps the world, is no more."[163] Herbert Casson greatly overexaggerated the subject as he portrayed it in his book, *The Romance of Steel*: "The five thousand workmen at Braddock were frantic with grief [King's Handbook for 1891 to 1892 gives an estimated total employment of 3,500 for the works.] . . . Carnegie, looking upon poor Jones as he lay in the hospital, sobbed like a child [*The Pittsburgh Commercial Gazette* indicated that Carnegie was traveling back to New York that Saturday, and even the family was not at the hospital when Jones died Friday night]. Ten thousand wet-eyed men marched with him to his grave [*American Manufacturer and Iron World* of October 4 stated that about five thousand were in attendance], and today the veteran steel-maker's most precious memory is: 'I worked with Bill Jones.'"[164]

There is absolutely no doubt that Jones was a very popular man. He was a member of the American Society of Mechanical Engineers, American Institute of Mining Engineers, Engineers' Society of Western Pennsylvania, and the Iron and Steel Institute of Great Britain, all without benefit of formal education and all of which published extensive obituaries extolling the deep respect that the industry held for him. In addition, he was active in the Grand Army of the Republic, Odd Fellows, and Freemasons. He had recently been championed for personally taking hundreds of men from the Pittsburgh area earlier that year to lead them in a relief effort. Jones and his complement of men had arrived only hours after learning of the disaster and provided aid to the people of Johnstown who were suffering from the immediate effects of the infamous flood, caused when the South Fork Dam collapsed on a branch of the Little Conemaugh River, following days of heavy rain.[165] His obituary in the *Transactions of the American Society of Mechanical Engineers*, a small part of which is included here, was quite heartwarming:

> Captain Jones was possessed of great physical strength and an indomitable will, but overmastering all, a most generous nature, and a heart as tender as any woman's. While quick of temper, he was ever ready to acknowledge and repair a mistake . . . was beloved by all who knew him. The men under his management worshipped him, and the community in which he lived, honored and respected him. The world is better for his life, but many hearts are made desolate by his death. If ever a man existed who was absolutely honest in every fibre of his being, such a man was William Richard Jones.[166]

An appropriate reminder of Jones's life might be one from the works of John Donne from his devotions:

> No man is an island, entire of itself; every man is a piece of the continent, a part of the main; if a clod be washed away by the sea, Europe is the less, as well as if a promontory were, as well as if a manor of thy friend's or of thine own

> were; any man's death diminishes me, because I am involved in mankind, and therefore never send to know for whom the bell tolls; it tolls for thee.[167]

The loss of Jones was immense. He did not leave this earth without performing one final service for Carnegie. He had cultivated a successor, and his name was Charlie Schwab.

From: Braddock's Field by Frank Cowan, 1878[168]

There the turning converter, while roaring with flame,
Pours out cascades of comets and showers of stars,
While the pulpit-boy, goggled looks into the same—
Thinking little of Braddock and nothing of Mars . . .

There the ingot aglow is drawn out to a rail.
While the coffee-mill crusher bombs, rattles and groans,
And the water-boy hurries along with his pail,
Saying Braddock be blowed! He's a slouch to Bill Jones!

Chapter Six

Pittsburgh Bessemer Steel Company and Open Hearth Steelmaking

The Panic of 1873 put one of Kloman's investments in an iron ore mine at risk. This Lake Superior mine, which supplied Lucy Furnace, was a part of Kloman's personal financial ventures. They profited Kloman alone and were not connected to the new Edgar Thomson Steel Company.[1] The financial depression was the result of the failure of one of America's most respected banking companies, Jay Cooke and Company. Cooke's failure was caused by its investment in and mortgage of lands that were grants made to the Northern Pacific Railroad. The estimated worth of the land was extremely overvalued by the bank. Its collapse led to the demise of several New York banks, which ultimately resulted in the closure of the Stock Exchange in New York for ten days.[2] The mine's failure during the Panic made the new steel company a target since Kloman was a partner. So the steel company and all of its investors were also put at risk to his creditors. Consequently, Carnegie coerced Kloman into relinquishing his shares in the firm, at that time known as Carnegie, McCandless and Company, until he could settle his affairs, after which time he would be reinstated. Meanwhile, he would receive a substantial yearly salary. A committee appraised the value of his interest in the company that Kloman accepted. Then, Carnegie bought him out. A disagreement over the interpretation of the details of this arrangement was later the source of Kloman's exit from the firm and his subsequent vendetta against Carnegie.[3]

Coincident to the many developments with respect to steelmaking at Edgar Thomson in the late 1870s, a different kind of revolution was being orchestrated by Kloman at Homestead, located across the river and about two miles downstream from Braddock, yet still within sight of the ET plant. The Pittsburgh Bessemer Steel Company (PBSCO) was planned "to be" Andrew Kloman's dramatic magnum opus of revenge against Carnegie, his Shakespearian plot. For Kloman, though, it was "not to be" because he died before the facility was completed. Yet in an ironic reversal of the plan, it was "to be" for Carnegie, since he later purchased this plant that became the flagship operation of the Carnegie Steel Company. For various reasons, under its new name, the Homestead Works became the largest and perhaps the most famous steel mill in the world.

Kloman had begun his attack on Carnegie in 1879 through the exploitation of his own mechanical genius to construct two new powerful rail and billet mills. In what might seem to be perfect timing for Kloman, a combination of events lent favor to Kloman's project. Carnegie had upset a number of Pittsburgh's smaller steel producers and/or rolling mill owners, who did not have access to enough steel to fully maintain their own operations.

Fantastic increases in the productivity and output of the Bessemer furnaces at ET, due in large part to ideas promulgated by Jones, allowed ET to far exceed the capacity of the plant's original rolling mill to convert all of the steel made there into rails. Jones developed the means of altering output in order to shift excess capacity from those furnaces to produce grades of steel that would not normally have been produced in a converter; for example, steel grades that were acceptable for axles, bars, beams and plate, and others.[4] This was not a small achievement. Jones then took advantage of the ability of the plant to roll this excess production of new grades of steel to make billets. This would not overtax the limited capacity of the rail mill, because it was located downstream from the billet mill. The additional semifinished, steel billet product from ET was then sold on the open market, so others would then roll it into the final shapes desired. Some local steelmakers felt that Carnegie had made them reliant on the surplus from Edgar Thomson, which he then promptly removed from the market after a new higher-capacity rail mill had been completed at the plant.[5]

Through Kloman's urging, a consortium of some local iron and steel makers was formed consisting of Kloman, W. G. Park, C. G. and C. C. Hussey, William Singer, Reuben Miller, and William Clark to build a Bessemer steel plant at Homestead (in Munhall actually). This new organization was called the Pittsburgh Bessemer Steel Company, and using Kloman's new rolling mills,[6] it could compete with Carnegie's Edgar Thomson Works. All of these men owned or were partners in firms that manufactured wrought iron or steel. The Husseys were recognized for the fact that their company was the first in America to make crucible steel that was routinely equal in quality to the English product. Within a few years Park followed suit.[7] He and his brother owned the Black Diamond Steel Works. Their plant was cited as being the largest crucible steel works in the world.[8] The elder Park, James, was also one of the principals in the Kelly Process Company, together with Ward, Lyon, Durfee, and Morrell, that in 1863, as previously noted, resulted in their control of the Kelly patents and essentially the process of Bessemer steelmaking in the United States.[9] Park personally did not make Bessemer steel, nor did the Black Diamond firm, but Black Diamond was a proponent of and also an early leader in the use of the open hearth process in this country. So Carnegie now faced formidable competition, at least on paper.

Andrew Kloman's hostility toward Carnegie was the last in a long history of divisiveness between Kloman and his business partners. In fact, his initial act of vengeance appears to have been directed toward his own brother, Anthony or Anton, whom Kloman believed drank too much, and in addition, Kloman felt that his poor work ethic not only directly affected the output of the forge on Girty's Run in Millvale but also negatively impacted the attitudes of the rest of the company's employees. This erosion of confidence in his brother began some years before but escalated with the introduction of partners from outside of the family.

Thomas Miller's association with the forge came about through his role as a purchasing agent for the Pittsburgh, Fort Wayne and Chicago Railroad. He procured goods for the railroad, including the well-regarded axles manufactured by Kloman.[10] Kloman needed capital to expand his company that Miller provided in return for one-third of the business. Miller, wanting to avoid the stigma of impropriety because of his position with the railroad, desired to be a silent partner.

Henry (Harry) Phipps's introduction to the forge's operations came about through his having been Miller's representative there. He also served as the firm's bookkeeper, through which he played a vital role. That role allowed Phipps, with Miller's help, to graduate from the clerical ranks into the hierarchy of the business world as an owner.[11]

Anthony's ouster was accomplished when Andrew Kloman convinced Miller to buy out his brother's share of the business for $20,000. After that distasteful affair was out of the way, Kloman realized that he was a minority owner in his own business, a very disturbing thought, so he began to focus his frustration on Phipps, who was a more available target, since Phipps had to perform daily bookkeeping duties at the plant while Miller did not. Phipps complained to Miller about Kloman's mistreatment of him, and eventually Miller agreed to grant Kloman one-half of Anthony's former share, giving Kloman a total of 50 percent ownership of the company. Having gained that position, Kloman wanted to be in complete control, and he began demanding that Phipps sell him his portion of the business. Phipps refused, and again Miller defended him. It was at this time that Kloman also began his attack on Miller. Miller's role became a public one with his purchase of Anthony's portion of the enterprise. Kloman accelerated the assault by having a notice published in one of the local newspapers stating that Miller was not a member of the firm and could not act as such. In a somewhat inconceivable turn of events, Phipps, who was typically meek, sided with Kloman when Miller approached him for support, even though it was Miller who had gotten him into the business in the first place and had protected him from Kloman. Enter Andrew Carnegie. Carnegie's introduction to Kloman was at the behest of his friend Miller, whom Carnegie knew in addition to Phipps from his childhood neighborhood when he lived on Rebecca Street in Allegheny, now a part of Pittsburgh's North Side. These relationships were nurtured because Carnegie's mother, Margaret, had occasionally done cobbler work for Phipps's father. Miller asked Carnegie to act as an arbiter in his dispute with Kloman at the forge, a hostile situation that now also included Phipps.

Carnegie met with each of the parties individually and finally as a group. It was a long and difficult process, almost impossible to resolve, because of the bickering brought about by the perplexing amalgam of hard feelings that Miller, Kloman, and Phipps had been harboring for each other. Ultimately, he had to rule against his friend Miller because Kloman and Phipps were the partners who actively ran the business. He suggested that Miller again become a silent partner. He made sure that Kloman resolved the issue of Miller's public standing in the company, made dubious by the libelous posting that Kloman had inserted in the newspaper. The settlement demanded that all of the parties were required to sign an agreement. It included a clause by which the silent partner (Miller) could be bought out forcibly after a six-month notice from the other partners. With the inclusion of that clause, Carnegie added an addendum that stated: "In the event of Miller's ejecture one-half of this interest would fall to my brother."[12] As might be expected, Miller was duly ejected, after which he and the elder Carnegie formed a competing firm, the Cyclops Iron Company, on 33rd Street, four blocks away from Kloman and Phipps Company's other mill (The Iron City Forge) on 29th Street in Pittsburgh. These plants were located across the Allegheny River from the Girty's Run Forge in Millvale. When the installation of the foundations for the new Cyclops mill was completed, they were found to be poorly designed and constructed, and despite the objections of Miller, Carnegie sought Kloman's help in addressing the problem. The net result was the merger of these two firms to form the Upper and Lower Union Mills. Since no one could find a way in which to compel Miller to work with Phipps, again, he was forced out of this new company, too.[13] It was at this time that Tom Carnegie became a partner in the firm of Kloman and Phipps.

This sequence of events is significant not only because it describes the way in which Andrew Carnegie first made his not-so-grand entrance into the steel business, but also for its insight into the demeanor of Kloman. The final confrontation between Carnegie and Kloman occurred at the Braddock Works. Carnegie insisted that before he would reinstate Kloman as a partner, following his involvement with the ore mine fiasco that put the new Edgar Thomson Steel Company in peril, that Kloman must first sign a letter of agreement, stating, "should he ever again engage in outside business investments his interest would revert to the other partners."[14] To say the least, Kloman was outraged and refused to comply, culminating in their final separation. Although this may appear to have been a preposterous affront on the part of Carnegie, its basis can be found in a far earlier agreement, when Kloman demanded that both his brother and Phipps sign, affirming that they "not engage in any other business whatever" (Kloman and Company, 1862[15]). Another ironic twist to a story that is overflowing with irony, later a similar "iron-clad agreement" was instituted at the Carnegie Company at Phipps's insistence. Its origin seems to stem from Kloman's earlier version and was an important element that led in 1901 to the end of the Carnegie Steel Company.

While Kloman could apply his mechanical and inventive genius to build and equip an impressive new rolling mill, he could not do so with the Bessemer plant. The Bessemer Steel Company (formerly the Pneumatic Steel Association) was protected by patents that it controlled. The Bessemer Steel Association also controlled the rail pools.[16] An important member of that syndicate was Andrew Carnegie, and it was extremely unlikely that he would have allowed Kloman to participate in the pool so that he could freely work in direct competition against the Edgar Thomson Works.[17]

The new converter plant at Homestead was built along the lines of the English furnace layout—two four-ton vessels face to face across a circular furnace pit (see figure 6.1 for a plan of the layout). Paul Krause in his *The Battle for Homestead 1880–1892* makes a compelling argument that it was in fact Alexander Holley (the architect of the ET Works) who designed the new plant's Bessemer shop and the blooming mill. Holley understood the process in sufficient depth and detail to know where the new mill could and could not legitimately circumvent his own as well as other patents. Later, Mark Brown in his PhD dissertation cast some doubt on this, but it still remains likely.[18] In many respects the return to English practice together with its unique new rolling mill design put the Homestead plant at somewhat of a disadvantage, especially when taking into consideration the character of the local labor force. To further complicate the matter, Andrew Kloman died suddenly on December 19, 1880, at only fifty-three years of age, not long before the converters made their first blow at Homestead on March 19, 1881. Holley died just short of age fifty on January 29, 1882, shortly after production began. That fact stymied the ability to confirm Krause's hypothesis that Holley designed

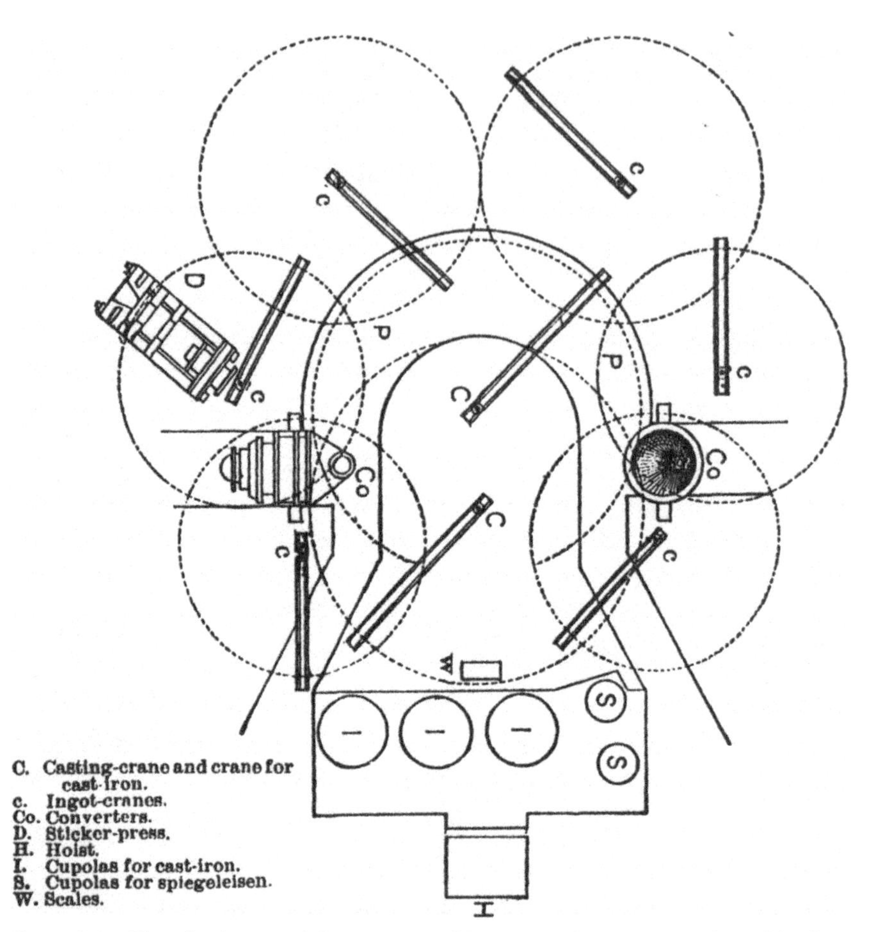

Figure 6.1. The plan layout of the Homestead Bessemer Shop which harkened back to the English plan where the converter vessels were directly opposite each other around the circular pit.

Howe, The Metallurgy of Steel, 1904.

the plant. No one else has ever been credited with it.[19] Kloman's demise left physical control of the new facility in the hands of his partner, William Clark.

Choosing Clark as the plant's new leader further frustrated operations. Clark was vehemently antiunion.[20] Unfortunately, the plant was located in one of the most firmly established bastions of labor unionism in this country, the Pittsburgh area in general and the town of Homestead in particular. The earliest documented organization of ironworkers in America, the United Sons of Vulcan (i.e., the puddlers' union) was formed in Pittsburgh in 1858.[21] In 1876 it was federated into the Amalgamated Association of Iron and Steel

Workers of the United States (AAISW).[22] For many years it was headquartered in Pittsburgh.[23] The *National Labor Tribune* was published in Pittsburgh and often served as the mouthpiece for a number of these local labor organizations, including the miners and coke workers, in addition to the iron and steel workers. It gave them a national platform from which to promote their ideals.[24] The union essentially controlled the entire political structure, as well as the police force in the town of Homestead. With this combination of circumstances, where the union controlled the town and management had command of the largest workplace, it formed the basis of serious antagonism that finally led to the immense tragedy of 1892, after being further stimulated by a number of different events in the intervening years.[25]

THE OPEN HEARTH PROCESS OF MAKING STEEL

As previously indicated, steel is an alloy of iron and carbon. This is also true of cast iron and wrought iron. We've further established that steel has less carbon than cast iron, but usually more than wrought iron, and that cast iron tends to be too brittle while wrought iron is typically too soft. Without addressing the fact that steel (iron) can be alloyed with materials other than carbon such as nickel, chrome, molybdenum, and others to enhance certain other characteristics of the metal, the most general way by which we can define steel is: a low carbon alloy of iron that most adequately meets the purpose of its intended use. The best rule-of-thumb interpretation that I have gotten from steel plant metallurgists is that, if the metal contains about two hundred carbon (mill jargon indicating 2.00 percent carbon) or less, it is steel; if it has more, it is iron. That definition is not without precedent, due to a higher potential for the graphitic form of carbon (graphite is what we use for pencil lead) to exist as inclusions in the metal matrix near levels of 2.00 percent or higher (see the phase diagram in figure 1.2). These inclusions would weaken the final product.

Prior to the introduction of the open hearth process, each of the early steelmaking methods was burdened with its own particular set of disadvantages. For instance, wootz could only be made in tiny quantities and required a special ore having very limited availability. Wrought iron could be made into steel in a puddling furnace, but that demanded a very restrictive specialized knowledge "in arresting the process at the proper stage of decarburization"; that is, by the educated guess of a very small set of individuals.[26] Also, both the wrought irons and wrought steels made this way were contaminated with slag. High-quality cast or crucible steel could only be made in small hundred-pound lots or less, and the source of raw material to make the steel was wrought (puddled) iron and/or blister (cemented) steel. Pneumatic steelmaking proceeded quickly, and the refined iron was made into steel in sufficient quantities, yet

> the chief limitation to the Bessemer process is in the entire absence of power to manipulate beyond very narrow limits, for the reason that, when a heat is blown, the sources of heat are exhausted, and the steel must be put into ingots before it becomes too cold. If it be overblown, the excess of [oxides] cannot be eliminated; if it be underblown, or if it be a cold heat, there is no means of increasing the temperature; if it be "wild," sufficient time cannot be given to allow it to settle, for it cannot be "killed"; and, if steel be not practically killed in the melting, the resulting ingots will be as porous as a sponge—honeycombed. Notwithstanding, the [Bessemer] process is simple, efficient in good hands and very cheap; it has its great field of usefulness and is here to stay.[27]

Nonetheless, despite all of these limitations, after a bit of coaxing, steel produced in an open hearth would adequately meet these challenges.

Carl Wilhelm Siemens, usually referred to as William but sometimes as Sir William, or Karl, or K. W., or even Charles William, immigrated to England from Hanover, Germany.[28] During a period in the early 1850s, William began experiments in an attempt to recover some of the heat lost by reverberatory furnaces. Together with his brother Frederick, he had studied a related subject during the 1840s. At that time it concerned the thermal efficiency of steam engines. Based on suggestions from his younger brother, who remained a German citizen, the pair developed a large heat sink comprised of firebricks that were loosely stacked in such a way that they provided a myriad of serpentine pathways by which the hot waste gas from the furnace could pass through the labyrinth with little resistance on its way to the chimney. In turn during this journey, most of its heat was transferred to the large volume of bricks. Bricklayers would say that the manner or style in which the bricks were piled was similar to an *open Flemish bond*.[29] Setting the bricks in this fashion presented a large surface area for the exiting gas to heat, making it easier to give up its accumulated warmth. The idea, which became known as a *regenerator*, was sheer brilliance, but in the beginning it had few practical applications. This type of pile of bricks was often referred to as *checkers* or *checker work*. Initially, regenerators were used in conjunction with solid fuels for steel reheating furnaces, as well as by comparable industries using glass and copper melting furnaces. This was somewhat similar to the earlier concept of burning excess gas generated from smelting iron and using it to warm the air in blast furnace stoves.

When burning fuel for heating a reverberatory furnace, not all of the heat energy can be exploited. In order to capture most of this untapped potential from the spent exiting gas (the *waste heat*), the waste gas was used to heat the bricks in the regenerator. Then that stored energy was used to raise the temperature of the incoming combustion air, transforming it by passing the air over the newly warmed checker bricks, before using the heated air to burn the fuel. Doing so would subsequently increase the total temperature of the furnace through its additive effect; that is, the total temperature of the furnace was raised by the measure of heat that the combustion air was increased above the ambient (outside) temperature after passing through the regenerator. Two of its important side benefits were a significant fuel reduction and the unexpected use of lower-quality fuels. Another totally unanticipated bonus arising from the invention came in 1857, when Edward Cowper proposed and later built a successful design of blast furnace stove that bears his name. He adapted the concept of Siemens's firebrick regenerator to his stove (discussed in chapters 4 and 5). His invention, still in use to this day, was opposed to that of the cast iron piped stoves, where blast furnace gas or other fuel was continually burned on the outside of pipes as the air passed through and was heated on the inside of the pipes, which had been the generally accepted method.[30]

On an early Siemens reheating furnace there was typically a single combustion chamber with two regenerators and two furnace chambers, which were adjacent but separated. Air for combustion was warmed as it passed first through one regenerator and then through one of the two furnace chambers, after which it entered the firebox to burn the fuel. The flame and hot gasses from combustion then traveled into the second furnace chamber to heat the glass, steel, or other material and passed out through the bricks of the second regenerator chamber, which were heated by the still relatively hot waste gas in the process, and then proceeded out the stack.[31] When the first checker chamber cooled to the minimum desired temperature, the flow through the system was reversed by means of valves. This *reversing* procedure was repeated continually, typically about every half-hour or hour, for as long as the furnace was on line. However, this system was impractical and virtually impossible to use when it was applied to a furnace that had only one heating chamber, such as a puddling or air furnace that required constant firing to maintain a high enough temperature. Achieving that goal required further innovation, but at first the obstacles to success seemed insurmountable.[32]

To accomplish this feat, the Siemens brothers found that it was vital to convert what became known as the *Siemens regenerative furnace* to use gaseous rather than solid fuels, before it became feasible for purposes other than reheating in the steel business.[33] Also plugging problems within the regenerator had to be addressed. In the past, solid fuel use brought fine particles of flue dust, as well as burned and unburned fuel, over with the flame, impacting the regenerator brick, where it would collect and plug the gas passages after numerous operating cycles. So Siemens chose a preexisting gas producer design to adapt his furnace to using gas. Initially, it was found that some minor modifications were required so that it would work correctly with his application. "There are two distinct principles embodied in the Siemens furnace, namely, the application of gaseous fuel, and the regeneration of heat by means of piles of bricks alternately passed over by the waste gasses and by the gasses entering the furnace before their combustion."[34] Produced gas complicated the furnace arrangement, because it required the addition of two more regenerators to heat the gaseous fuel before combustion, bringing the total number on the furnace to four. Manufactured gas of this type made from coal usually contained ammonia, tar, and other hydrocarbons that could condense where it met the relatively cooler checker firebrick, or through mechanical impingement, and it would coat the regenerator. The hydrocarbon coating could then be coked from the high temperature as the brick was reheated. Switching the gas chamber to air for reheating also reduced deposits because the waste gas still contained a small amount of oxygen, causing them to burn off. A condensing chamber located after the gas producer but before the gas regenerator was later introduced to reduce deposition of hydrocarbons by collecting and draining them while still in the liquid state. Due to the low heating value per cubic foot of the manufactured gas, huge volumes of this fuel were necessary to heat a furnace to the high temperatures required to melt steel. This was the major reason it was necessary to heat the gas in addition to heating the air in a regenerator. Otherwise, injecting the enormous mass of the relatively cold producer gas necessary for elevating the contents of the furnace chamber to several thousand degrees Fahrenheit could have significantly lowered the temperature of the entire furnace. In addition, all of the free heat from the waste gas would have been lost, disappearing up the chimney and into the atmosphere. Slack (bituminous or soft) coal was most frequently used to fuel the producer, but more inexpensive, very low-rank fuels such as lignite, peat, coke breeze (the fines from sizing or screening coke that was normally a waste product), and even wood or sawdust was often used to make gas.[35]

There were several different iterations of the plan executed during the evolution of furnace design, and a number of patent applications were filed, starting with brother Frederick's patent of the regenerator.[36] By 1862 William Siemens had succeeded in creating a rotating drum furnace capable of making wrought iron or steel directly from ore. This had been one of his goals.[37] Although he apparently could routinely make steel, the rotating drum was not deemed to be a success because the steel suffered from sulfur contamination (hot shortness).[38] So he moved on, adapting the regenerator technology for heating a conventional reverberatory or puddling furnace. The historic significance of these developments should not be underestimated, because by the application of the brothers' enhancements, Siemens's furnace was capable of maintaining the iron or steel in the molten state throughout the process.[39] This is the most important benefit of his invention. The added heat gained when the combustion air passed through the furnace regenerators made it capable of maintaining the temperature at the hearth above the approximate 1535°C (again, refer to the phase diagram in figure 1.2) required to prevent solid bits of metal from precipitating out of the bath as it became more pure, exactly the goal of the old puddling methods. Furthermore, this was true regardless of the carbon content of the molten iron. Siemens's new practice of introducing the gas and air alternately via one end of the furnace chamber and then the other (reversing) had the beneficial effect of decreasing the temperature differential across the furnace. Previous furnace designs usually had a hot end and a cold end (relatively speaking), because as the hot flame entered at firebox end, it gave up its heat to the material in the furnace chamber, but the flame became depleted of its energy and was at a considerably lower temperature as it exited the opposite side and went up the stack. Alternately burning the fuel at either end eliminated this discrepancy for all practical purposes.

It seemed that Siemens accomplished his goal of making steel directly from ore through his development of the rotary drum furnace (albeit contaminated with sulfur), and he furthered this concept with the regenerative puddling furnace. While it was proved that both methods could be suitable for making steel, the statement that it was a "direct process" is not totally true. This was due to the fact that he always needed to use previously manufactured pig iron from a blast furnace in order to reduce the oxygen in the ore through the mechanism of using the carbon contained in the pig. To this day for a multiplicity of reasons, steel is not being made directly from ore to any great extent. Notwithstanding this, in 1864 John Percy expressed his opinion of exactly how significant these concepts were when he said, "I anticipate results of great importance in Metallurgy from the extension of Siemens's furnace, which appears to me to be founded on truly philosophical principles."[40]

In 1864 Emile Martin and his son Pierre, of Sireuil, France,[41] began using a Siemens regenerative furnace under license to experiment with making steel. There is a bit of confusion with this development being credited to Pierre alone; his middle name was also Emile, or the pair are often thought to be brothers. They devised a method of making steel described as dissolving wrought iron in a bath of molten pig iron. The famous French metallurgist and philosopher Reaumur had proposed this idea in 1722.[42] Later they successfully substituted steel scrap in the place of the wrought iron.[43] In 1867 the Martins were awarded a gold medal at an International Exhibition in Paris for samples of the steel made by their new process.[44] An important aspect of their process was the ability to convert a substantial portion of old wrought iron rails into steel. It was at about this time, due to their recognition in Paris as well as the growing attention paid to Siemens's new design of furnace, that the first utterances emerge in contemporary literature, describing this new way of producing steel as being converted or made on the open hearth of a Siemens regenerative furnace, or more simply, what we typically call the *open hearth process*.[45]

Siemens's metallurgical ideas and inventions were contemporary to those of Bessemer. The success that both men had with their discoveries was accompanied by the inability of both of these inventors to create substantially successful enterprises without the subsequent innovation and intervention of outsiders beyond either man's sphere of influence—Bessemer had Mushet, while Siemens had the Martins.

The fact that we had two separate modern methods for creating mass-produced steel—the open hearth and the Bessemer converter both materializing at the same point in history—seems truly extraordinary. The two processes seem to be diametrically opposed. That is:

- one was slow, the other fast
- one produced good quality and the second acceptable
- one required skill and understanding to facilitate its operation while the other needed little of either

The open hearth, of course, was the slow, arduous, elusive, scientific path to higher quality. It wasn't the best steel when compared to that from a crucible, but it could create much better grades of material than a Bessemer converter for use as boiler plate and tires for locomotive wheels in high volume, something not possible from a crucible. Of course, you could make large amounts of bad steel equally well if the process was not adequately controlled. While the Bessemer process seemed to leap into the forefront almost immediately after a brief false start, adopting the use of the open hearth was slow and plodding, taking more than a decade to work through the major bugs before starting to more firmly establish itself.

The *Siemens* open hearth process, or as Siemens called it himself the "fusion of steel on the open bed,"[46] had its basis in his direct steelmaking experiments. "His object was to make steel direct from the raw ores, without the intermediate use of huge blast furnaces and laborious refining processes."[47] But, according to W. T. Jeans, steel made by his methods, normally called the *Siemens process*, wasn't developed by William Siemens until 1867.[48] The product made by the Martins, however, was devised between 1864 and 1867,[49] before Siemens, during experiments using a furnace licensed from Siemens and originally called the *Martin process*. Now it is usually called the *Siemens-Martin process* or even the *pig and scrap process* due to its admixture of those two raw materials. Discussion of this subject is sometimes complicated by the controversy that arose between the Martins and Siemens over the ownership rights to the process, which was somewhat similar to the disagreements that occurred between Bessemer and Kelly or Bessemer and Mushet. Siemens was awarded the glory of invention for the furnace even though the Martins' practice of making steel in it predated Siemens's pig and ore process, and despite Siemens's earlier failure with the "direct process" in the rotary drum.

The Martins' process used oxygen from scrap steel or wrought iron to remove the carbon from pig iron. For all practical purposes it was the inverse of Siemens's later 1867 concept, the *pig and ore process*. He used oxygen bound to the iron in the ore that was added to a molten bath of pig iron in order to remove the carbon from the pig iron, and thus reduce that element to levels considered acceptable for steel. So while there may appear to be only a subtle or perhaps merely a semantic difference between the two processes, in actuality they differed materially as the scrap added to the chamber in the Siemens-Martin process had to first be exposed to the oxidizing influence of the furnace flame before it could react with the carbon in the molten iron. It was imperative that the scrap was oxidized (in essence partially burned). Otherwise, their system would not have been practical. Siemens's process was not the same as making steel from ore, what he was calling the direct process, nor is it the same as using the pig and scrap method. One used the addition of pig iron to oxidized scrap and the other used the addition of iron ore (oxidized iron) to a bath of molten pig to gain the same result. The remainder of the routine was more or less very similar.[50] "Hence the open hearth process is known as an oxidizing process."[51]

There were other steelmaking schemes being implemented and proposed. An 1867 article in the *Journal of the Franklin Institute* reveals a critical difference; for example, that: "Messrs. Martin however, have succeeded in making their process a commercial success." This article, "New Modes of Making Steel," made a comparison of the pig and scrap process to a corresponding development presented by Berard at the Paris exhibition. However, it suggested that the Martins could at least profit financially from their efforts and that their ideas were being adopted by some important French iron and steel companies. There was no mention of Siemens, other than for his furnace design.[52] Given the date of his development, it was not surprising that the Siemens process ended up being remarkably similar to that of the earlier so-called Siemens-Martin formula.

Like the puddling furnace that it mimicked, using an open hearth furnace to make steel was a laborious and technically challenging (for that time) undertaking for the men who were charged with its operation. A complete cycle or heat for making Siemens-Martin steel took four or five times longer than was required to puddle wrought iron, which itself took about ten times longer than a Bessemer to make a heat of steel. Since an open hearth furnace was nothing more than a puddling furnace that was equipped with Siemens regenerators and burned gaseous instead of solid fuels, it also demanded skill to recognize the nuances indicative of change in the furnace. This was especially true for that of the metal bath. Given the extensive knowledge for making wrought iron that a puddler possessed, it was highly unlikely that he could just step in and favorably meet the demands imposed upon him by the open hearth furnace, as that skill set was significantly different.

The *first helpers* (typical terminology in the United States but known as *first hand melters* in the United Kingdom) or the men in charge of an individual open hearth furnace's operation had to develop certain skills that, though often substantially different, plainly paralleled many of those executed by a puddler or a melter on a crucible furnace. One talent in particular required the ability to determine the correct temperature of the molten bath. A puddler was not charged with knowing this detail to any great degree of certainty because his furnace was not capable of maintaining the metal bath hot enough to remain molten throughout the process. In fact, as was previously explained, that was the "magic" of a puddling furnace's operation. A melter on a crucible furnace was usually able to judge the temperature with his eyes, occasionally meeting with failure. The penalty for any such misstep was the spoiling of sixty to a hundred pounds or so of steel. On early open hearth furnaces, whose capacities were usually in the range of three to forty tons, this same error in judgment was a hundred to a thousand times more costly. There were neither thermocouples nor other electronic devices on which to rely for making temperature measurements in this era. The furnace chamber was very bright and variable with flame, and the molten bath was covered with slag. Both interfered with visually estimating the temperature. A "ballpark" method of determining the heat of the metal bath, as described in the *Journal of the Iron and Steel Institute*, was presented as follows:

> When the workman desires to find if he has reached approximately the right temperature, he places in the furnace, through a hole in the working door, a round iron bar 0.31 inch in diameter. It is left in the furnace twenty-six seconds, measured by a rough pendulum beating seconds, and is then withdrawn quickly.

> If the furnace is very hot the end of the bar will be at a white welding heat, and when drawn quickly through the air will throw off sparks. If the temperature of the furnace is too low, the rod comes out of the furnace red or yellow.[53]

There were other methods of temperature "guesstimation,"[54] and numerous other instances where the good judgment of the furnace operator was required to satisfactorily move the heat forward through the process, such as:

- determination of carbon content
- the reducing or oxidizing nature of the atmosphere inside the furnace chamber
- assessing the basic or acidic nature of the slag
- analyzing the condition of the refractory walls and hearth, and more

These abilities made the first helper far more indispensible than a pulpit operator on a Bessemer converter, and it also endowed him (as well as his compatriots) with a much more favorable position for negotiation.

Initially, the Martins used several different furnaces in tandem to make a single heat of steel. Early in their odyssey, there was:

- one to preheat the scrap
- a separate one to melt the pig
- one to make the steel
- finally another for melting spiegel[55]

This quickly was transformed into properly sequencing the order of events in just a single furnace. The slowness of the operation was a benefit as well as a curse. The curse was that it significantly reduced furnace output when compared to a converter. Some of the benefits were that it gave sufficient time to work the carbon out of the steel to a reasonably accurate chemical end point, and it reduced the constant rabbling of the bath that was so necessary in puddling. It also practically eliminated the largely unrecognized detrimental effects of nitrogen contamination that was prevalent in Bessemer steel. It does not mean that the open hearth eliminated the hellacious hot labor identified with making wrought iron, but it reduced much of the extended intense hard work to shorter spurts. It also introduced long spans of quiet time that often occurred during periods when the charge of pig iron or scrap was being melted or while waiting for the chemistry of the bath to show a change.

Making a heat of Siemens-Martin steel during the early history of the process roughly followed the accompanying simplified procedure, although there was some variation for numerous reasons:

- Charge pig iron into the furnace (it could have been premelted or preheated or put on the hearth cold to melt).
- Add scrap (either preheated or cold) in such a way and in a number of different stages so as to be exposed to the oxidizing effects of the flame before melting and blending into the molten pig. After a short time this was accompanied by a visible reaction in the furnace where the oxygen from the heated scrap combined with the carbon in the pig iron to release gasses that made the liquid bath appear to be boiling. Old texts refer to this as *ebullition* (approximately four to six hours elapsed to this point).
- Begin to routinely take samples from the bath using a sample ladle to pour a small test ingot, quench it in water to cool, and do one or both of the following with the ingot: break it with a sledgehammer and visually determine the carbon content from the appearance of the fracture, and/or hammer the ingot flat and bend it back on itself to determine its tenacity and therefore the percentage of carbon it contained.
- Make adjustments to the chemistry through the addition of more scrap, or on occasion iron ore was used (more modern). The direct oxidizing influence of the flame also had an effect on the bath; that is, contact of the flame and the exposed molten metal.
- Decarburize completely, and then add the appropriate amount of cold spiegel, placed on the slopes of the hearth to melt and dissolve into the bath (but could be added when molten instead) to bring the steel to the desired carbon content. Then wait about twenty to thirty minutes and tap the furnace into a ladle by removing the sand or clay plug from the tap hole on the outside and use a long pointed bar to poke it.

By comparison, creating a heat of steel by Siemens's pig and ore method was described as follows:

> The charge consists mainly or entirely of pig-iron, which is placed on the bottom and round the sides of the furnace. Melting requires four or five hours: then ore of pure character is charged cold into the bath, at first in quantities of four to five hundredweight at a time. Immediately [after] this is done a violent ebullition takes place; and when this has abated, a new supply of ore is thrown in—the object being to keep up uniform ebullition. Care is taken that the temperature of the furnace is maintained so as to keep the bath of metal and slag sufficiently fluid; but after the lapse of some time, when the ore is thoroughly heated and reduction is taking place rapidly, the gas may be in part shut off the furnace, the combustion of carbon in the bath itself keeping up the temperature. In the course of the operation the quantity of ore charged is gradually reduced, and samples are taken from time to time of both slag and metal. When these are satisfactory, spiegeleisen or ferro-manganese are added and the charge is cast.[56]

One assessment of the proper mixture of materials used in making a heat of steel by the pig and scrap method around this time called for 9 to 10 percent pig iron, about 6 percent spiegel, and the remainder scrap steel or scrap iron.[57] These proportions could vary widely, because the deviations frequently were based on locality and numerous other circumstances.

> For a 3 ton charge, about 6 cwt. [hundredweight or 100 pounds] of grey pig-iron is introduced into the furnace already heated to whiteness; heavy scrap is placed on the bed of the furnace near the bridges, where it is thus heated to whiteness before introduction into the bath of molten iron, while rails about 6 feet in length, bars of iron, etc., are introduced into the shoot [chute] previously mentioned, their lower ends only resting upon the hearth or bed of the furnace; the bars descending as their lower extremities melt away, when other bars are added to replace them until the whole charge has been introduced. After about four hours from the commencement of charging, during which time an oxidizing flame has been maintained, and the gas and air valves reversed after every twenty minutes, the whole charge is melted. Samples of the metal are now taken out in an iron ladle from time to time, and cooled in water; and when this test specimen exhibits on breaking the necessary toughness and malleability, notwithstanding its sudden cooling in water, from 3 to 4 cwts. of spiegeleisen, according to the temper required in the steel, is charged through the side doors on to the hearth of the furnace close to the bridge at the flue end of the bed; in about twenty minutes it will have melted and run down into the bath of melted metal; the fluid mass is then well stirred so as thoroughly to mix the charge, which is then tapped as quickly as possible to prevent the loss of manganese in the form of slag. To tap the metal, the sand near the tap-hole is removed and the hard crust formed on the surface bottom is pierced by a bar, on the withdrawal of which the metal runs out into a ladle, furnished with a plug, similar to those used in the Bessemer process; from which it is run into ingot moulds in the usual manner. The process from beginning to end occupies about ten hours.[58]

For an idea of what an open hearth operator would have done after removing a sample of steel from a test ladle: he would flatten it with a hammer, bend it back to break it, and then observe the fracture.

> This shows, at an early stage, white, like refined iron, becoming afterwards grey and dull. At the next stage, the fracture assumes a steely appearance, being, at the same time, more or less brittle; and at the last stage it begins to bend and tear. When this latter point is arrived at, the percentage of carbon is very small . . . and it is time to restore a portion of it by means of spiegel-eisen . . . It melts rapidly, and after one or two more tests, the process is concluded, and the metal is tapped out . . . If the bath of metal is at this point allowed to remain too long before adding the spiegel-eisen, it rapidly becomes deteriorated. The crystallized structure of the test pieces shows that it has been attacked by oxygen (in

> some form), —in common phrase, that it is burnt; and experience proves that when once these crystals have begun to form, it is a very difficult task to remove them. The steel, in fact, is spoilt.[59]

Originally, the open hearth, as well as the Bessemer converter, was an acid process, because the refractory and/or sand lining of the furnace vessel was usually comprised of silica and exhibited acidic properties. Therefore, no phosphorus or sulfur could be removed during the refining of the metal. As previously noted, that was a major failing (sulfur contamination of the steel) of Siemens's rotary drum process for making steel directly from ore. The sulfur infiltration came from the metal's contact with the furnace's flame. The flame was contaminated with sulfur from the coal, because the sulfur was contained within the coal matrix from which the producer gas was made. The sulfur was released as a result of the coal's decomposition during the gas manufacturing process. So if steelmakers were not very careful when using the acid open hearth, they could corrupt the final product with sulfur from fuel or pig iron, as well as phosphorus from scrap, the pig, or numerous other sources. Since these adulterants were not destroyed, they were compounded roughly as the sum total of all the sources of contamination. Though there were some ways proposed to remove the contaminants via a slag, acid open hearth steelmaking required much diligence to create an acceptable end product in which only tolerable levels of these impurities were found.

In 1878 Sidney Gilchrist Thomas and his cousin Percy Gilchrist were credited with devising the basic (vs. acid) steelmaking process when they developed a basic refractory lining and the means to almost completely remove phosphorus from the steel. This was accomplished through the injection of a large amount of lime into the vessel of a basic lined Bessemer converter. Previously (as early as 1856), Bell, Snelus, and others suggested using either a basic lining or adding lime, but until applying the idea as concocted by Thomas, a practical concept for dephosphorizing mass-produced steel proved elusive.[60] The Gilchrist cousins developed a method of producing an acceptable quality, chemically basic, refractory brick for use in the lining of a converter by substituting boiled (dehydrated or water-free) tar as a binding agent for the brick's component materials prior to firing them in a kiln. This was a key aspect of their system.[61] Thomas died soon after the idea was deemed a success, and cousin Percy carried on the work of promoting the concept.

"The lining itself did nothing toward the removal of phosphorus; but it was the added basic material that effected the end result, and the question was how much of that basic material was requisite to remove a large amount of phosphorus from the iron."[62] Finding an acceptable material to create a basic lining, in addition to adding a base like lime to form a slag to remove phosphorus during the refining period, turned out to be the proper combination for success. Simply adding lime alone to an acid refractory vessel severely corroded the lining of the furnace rather quickly.

> When oxidized, phosphorus and sulphur form acid compounds, but this only takes place when they are exposed at a high heat to the action of lime, or some other active mineral base with a stronger affinity for them than they have for iron. As the use of lime in an acid-lined furnace, as a reducing agent, would flux and destroy the lining, it is necessary to have a non-acid lining, or what is known as a basic lining, to allow of the reactions taking place without much wear on the lining.[63]

Later, the remaining bits of sand that adhered to the pigs of sand-cast pig iron were identified as being responsible for the significant erosion of a basic furnace's lining due to the sand's acidic nature.[64] The phosphorus that was removed from the iron during refining was prevented from reverting or returning to the bath of molten metal as the chemistry changed by routinely removing the slag from the furnace and disposing of it. Since phosphorus contained in the slag was often bound with iron, removing this contaminant was a source of considerable loss in the final yield, especially when not handled properly. Because slag from British steelmaking plants contained large amounts of phosphorus, it was often used as fertilizer on farm fields near the furnaces.

CARNEGIE'S HOMESTEAD WORKS

In October 1883 the principals of the Pittsburgh Bessemer Steel Company approached Carnegie and proposed that he purchase the mill from them. Since Kloman was dead and Clark's tenure had been a failure, there seemed to be no viable leader on tap to lead their organization. They had already endured five different labor disputes at the mill in its short two-and-one-half year life, including one where the strikers had killed one man and assaulted others. A later strike blossomed into a nationwide labor dispute, involving the iron and steel industry.[65] Carnegie obliged the owners and bought them all out for $350,000[66] ($1.2 million according to Kenneth Warren[67]). Only William Singer remained. His payment came in the form of a stake in Carnegie Brothers, an offer that was made to all but refused by the others.[68] The purchase of this plant by Carnegie and the acceptance of stock in lieu of cash by Singer was probably one of the more prudent decisions that either man had made during his life.

Unlike the management of the former Pittsburgh Bessemer Steel Company, operations of Carnegie's Homestead Works were put under the able command of Julian Kennedy,[69] who was responsible for much of the triumph in the operation of the blast furnaces at ET. Under his tutelage, his new charge became a rapid success, although the union continued to be an impediment. By July 1886, it was producing more than ever.[70] In 1887, little more than three years after the plant's purchase, the converter department's output set a world record.[71]

Part of that success was due to Carnegie's insight in changing the focus of the facility from producing rails to making beams and plates, as well as overcharging the stated capacity of the converters.[72] The rolling mill at Homestead was more powerful, permitting the use of larger ingots and resulting in shorter pit delays due to faster teeming and from stripping and setting up fewer ingot moulds. Consequentially, this increased the capacity utilization of the converters; more tons and blows per vessel per day. This plan also took advantage of ET's history for creating rails while permitting a more efficient and full use of that facility.

The year 1886 proved to be pivotal for Andrew Carnegie. His brother Tom died from an ailment that began as a fever and ended in pneumonia. His death was ultimately a complication of alcoholism that was directly connected to the stress of having to be the ever-present face of the Carnegie Company, because his older brother, Andrew, was mostly absent. His mother, Margaret, died later that year from pneumonia, while Andy was sick and near death from typhoid. They had been together at their cottage in the Allegheny Mountains at Cresson, Pennsylvania. The usual story is that they decided to lower her coffin out the window in order to avoid carrying her body past Andy's bedroom door and chance disturbing the frightfully ill man. For some time they didn't even inform him that she had died. Freed from the care of his mother, Carnegie swore eternal allegiance to Louise Whitfield, whom he married the following year.[73] Also, during 1886, he embarked down a new path to make steel, for the Homestead Works at least, using open hearth furnaces.[74]

The *American Manufacturer* of April 23, 1886, reported that Carnegie, Phipps & Company had announced their intention to build a four-furnace open hearth facility at its Homestead plant. Also, a plate mill would be installed.[75] The first furnace went into operation on October 27 and the second on November 22 of that same year. Built of acid brick, they were designed by Horace Lash and Henry Aiken. According to Earnshaw Cook, each was of fifteen tons capacity (eventually totaling four of the same size).[76] However, the *American Manufacturer* stated they were supposed to be thirty tons apiece (thirty-five gross tons per the *Directory of Iron and Steel Works*, 1886).[77] The report of larger furnaces was supported by the fact that the central hydraulic cranes on the floor-plan layout of the shop in figures 6.3 and 6.4 were rated for forty tons, consistent with the required load rating for equipment to process a thirty-ton heat plus the weight of several tons for a ladle with a refractory brick lining. Cook states that "immediately following the operation of these furnaces experiments in the making of a basic hearth of dolomite and tar went forward."[78] So these first furnaces were smaller than initially reported, probably in order to provide more leeway on the learning curve for making this "new" type of steel, as well as to better accommodate the operators, giving them the ability to experiment more deliberately with the new facility so that they could accomplish the stated goal of "basic" steelmaking. In fact, according to a report made in the *Journal of the Iron and Steel Institute* of Great Britain regarding a visit to Homestead and other facilities in the United States during 1890, which was conducted more than three years after operations began, the author indicated that the furnace capacities in the original shop at Homestead were "two 15-ton, five 20-ton, one 35-ton,"[79] confirming Cook's report.

Open hearth furnaces of the Lash design initially had circular- or elliptical-shaped hearths (for example, see the two furnaces on the extreme left of the plan view in figure 6.3 or the structure in the right of the photo in figure 6.2), but soon they reverted back to a rectangular shape as is also shown in figure 6.3. Lash's configuration of the hearth was originally thought to facilitate the use of natural gas for fuel instead of producer gas that was typically employed. Using natural gas greatly simplified the arrangement of the furnaces at Homestead. Although with hindsight it may appear to be the obvious choice, substituting natural gas for producer gas was a brilliant stroke of genius for a number of reasons. Remember, the Pittsburgh region was one of the pioneers of the natural gas industry. Wells had to be drilled to satisfy the tremendous gas volumes demanded for heating a furnace. (And pipelines had to be laid from the gas fields located many miles distant from the plant.) This ultimately resulted in a simpler furnace, since an equivalent volume of natural gas has roughly seven to ten times the heating value of producer gas. So only about one-tenth the volume of producer gas was needed to heat the furnace. Reducing the volume by nine-tenths meant that it was no longer necessary to heat the fuel, thus eliminating the two regenerators normally required for preheating the gas. In fact, natural gas did not preheat well because it readily disassociated into its hydrogen and carbon constituents at furnace checker temperatures and would deposit carbon on the brick. This forced the removal of the gas regenerator because carbon deposition would quickly plug the checker chamber, rendering it ineffective, and that would shut down the furnace.[80]

Since no gas producers needed to be built when using natural gas, no capital investment was required. With nothing other than a simple pipeline:

- No labor was required to unload the huge amounts of coal needed to make the gas.

- No men were needed to operate the producers twenty-four hours per day.
- No mechanics were needed to maintain, repair, and replace the equipment.
- No one had to clean up, transport, and dispose of the great volume of ashes that were not generated.
- And most importantly, no payroll was needed to pay all of these nonexistent employees.

In addition, in March 1886 a new business unit was formed, named the Carnegie Natural Gas Company. Its purpose was to drill and own the wells, build the pipelines to distribute the gas to the various Carnegie companies, and to sell gas to others.[81] So eliminating the gas producers yielded a huge cost advantage, not from the capital investment, but from the never-ending operating expense side of the ledger. If the magnitude of these savings is doubted, a quotation from a pamphlet from the natural gas industry in Pittsburgh should set the issue straight about the benefits of using natural gas: "An honest mill man will tell you now that he could not afford to use coal if it were given to him on the ground without cost. Park Brothers & Co., estimate that their savings is so great that they would not be justified in taking coal free of charge, delivered to their property."[82] The savings were undoubtedly high, but the source of the information must be considered. Regardless, the merits of natural gas as the fuel of choice can simply be demonstrated on the basis of its selection, since the cost of coal in the region was only cents per ton (in large volumes). Also, consider that people used coal because it was so much more cost effective than wood, even though Pittsburgh was situated in the middle of a forest. The Carnegie Company's capital investment in gas properties, wells, and pipelines for Homestead was offset somewhat by the distribution of its cost over the entire company as well as the profits earned by the sale of this fuel to others. It should be understood that almost all of the larger Pittsburgh area steelmaking shops of this period made natural gas the sole source of their fuel.[83] Still, none of these shops compared in size to Carnegie, at least as reported in the 1886 *Directory* (see table 6.1). Natural gas obtained from the Haymaker Field near Murrysville, Pennsylvania, began to be used at the Edgar Thomson Works as early as 1883.[84]

The furnaces in the #1 shop were constructed at ground level, therefore the two regenerators for each furnace had to be built below grade. They were constructed integral to each of the long axis of the two waste heat channels to the chimney. A ladle pit, also below grade and about a foot below the bottom of the hearth, was provided at each furnace for tapping the molten metal upon completion of the heat. To give an idea of the size of this pit, a ladle large enough to hold forty-five tons of steel (much larger than required for the original furnaces) was slightly more than nine feet in diameter and seven feet in depth.[85] A combustion air passageway led from the ladle pit, then passed underneath the furnace to a three-way valve near the chimney; it served to preheat the air slightly, evacuate the smoke and gas from the ladle pit, and simultaneously cool the bottom of the furnace to help prevent this area from overheating.[86] The furnaces were arranged so that their backs (tapping sides) faced each other across the pit with the charging floors facing the outside walls of the shop. A centrally located pit provided room for two ladles (one for each furnace) and two teeming pits (common to both furnaces) and had a single large, hydraulically operated ladle crane that was positioned to serve each pair of furnaces (see figures 6.2, 6.3, and 6.4). In addition, there was a pair of five-ton ingot cranes and sixteen-ton auxiliary cranes for each individual furnace to use. These were also hydraulically actuated. "After overcoming many difficulties, Lash as superintendant began the successful manufacture of basic steel, the first commercial heat being tapped from No. 1 furnace on

Figure 6.2. This 1889 view of the Number 1 Open Hearth Shop at Homestead shows the pit area between the furnaces. The heavy square box girder hydraulic crane at left of center is the 40-ton capacity central ladle crane. To the right foreground is a 5-ton auxiliary crane (with what appears to be a furnace tapping spout laying at its base) and on the extreme left is a portion of a 16-ton ingot crane. Only one furnace is in the view at the extreme right of the photo. Note the dome shaped portion of the furnace structure beyond the downward sloping gas and air ports at the end of the furnace in the foreground. This would be the location of the hearth and is indicative of a Lash furnace design.

United States Naval Institute.

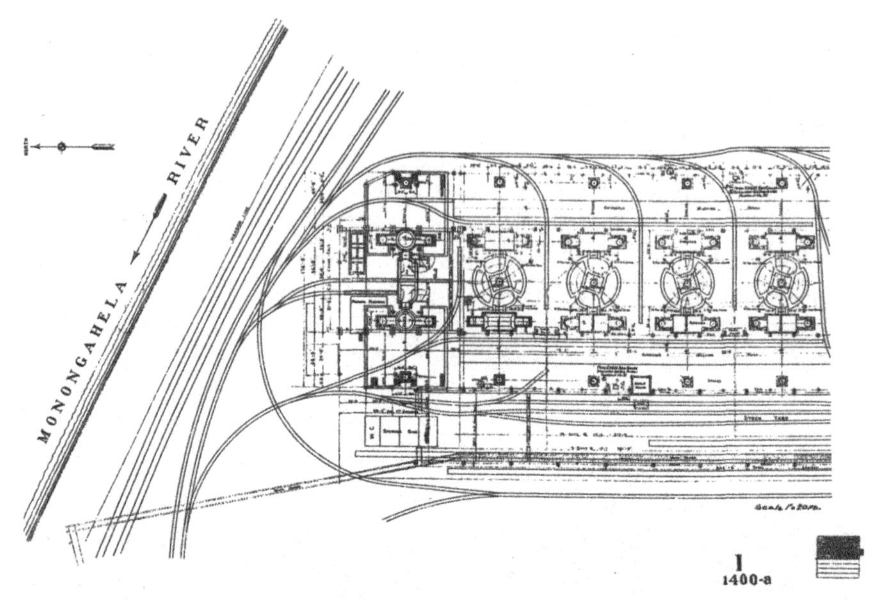

Figure 6.3. Plan view of the Number 1 Open Hearth Shop of the Carnegie Steel Company (USS era) circa c. 1927 or later (the last update noted on the drawing). Originally a four-furnace shop that was expanded to eight and shown as a ten-furnace shop in this rendering, this was considered to be the first commercially successful basic open hearth shop in the United States. The circular casting and teeming pits reminiscent of English Bessemer practice can be seen interposed between each of eight furnaces of the expanded Carnegie shop. The two furnaces (without the circular pit in between) at the extreme left, have furnace hearths with a circular shape, which is indicative of a Lash furnace design, although these two may be furnaces with moveable roofs that were known to exist at the Carnegie Company. The moveable roofs were used to facilitate recycling of pieces of scrap that were too large to fit through normal sized furnace doors and were charged through the top of the furnace.
Archives Service Center, University of Pittsburgh, William J. Gaughan collection.

March 28, 1888."[87] Having reached that milepost, the Homestead plant of Carnegie, Phipps & Company was granted the lofty honor of being named the first commercially successful basic steel operation in the United States.

This should not be interpreted to mean that it was at Homestead where the first basic steel was made in America. It was not. The first basic steel produced in this country was made using a basic-lined Bessemer converter (the process originally proposed by Thomas and Gilchrist) at the Bethlehem Iron Company on May 24, 1884.[88] From this it was determined that making basic steel in America by the Bessemer process was not prudent. The addition of large amounts of lime or a similar material needed to dephosphorize the metal required some additional source of heat to liquefy (melt) the lime, because it had to be added after silicon was removed in the blow (the primary source of fuel). The requisite heat for melting lime was generated through consuming (burning of) phosphorus contained in the iron during a *re-blow* or *after-blow* subsequent to the addition of lime.[89] Since pig irons produced in this country were not usually contaminated with phosphorus to the extent required (approximately 1.5 to 3 percent) to make this part of the operation feasible,[90] the use of the basic Bessemer process was too impractical to be adopted in the United States.

To further clarify the meaning of "first commercially successful," Homestead's was not the first basic open hearth steel produced in the United States, either. The first was made experimentally (not totally successfully) in one furnace at the Otis Iron and Steel Company in Cleveland, Ohio, between January 19 and April 6, 1886.[91] It took considerably longer to make steel at that time using a basic furnace, so the practice was quickly abandoned there, and the experimental furnace was converted back to the faster, more favorable, and profitable (for them) acid furnace configuration.[92]

Likewise, neither was the Homestead Works the first open hearth shop in Pittsburgh (actually Allegheny County because both Homestead and Munhall were just outside the city limits), nor did it have the greatest number of or largest capacity furnaces in operation. As reported in the 1886 *Directory of Iron and Steel Works*, there were twelve open hearth steelworks in operation at locations in Pittsburgh and Allegheny County (forty-two in the United States), which contained twenty-three furnaces, not including the four at Homestead (seventy-one listed for the United States). See table 6.1 for details.[93]

Granted, while it is extremely important to understand what Homestead was not, one should not use those facts to diminish the significance of what Homestead actually was. Although many, many people have excellent ideas, consistently producing practical results while simultaneously being capable of profiting financially is a cornerstone of the capitalist system in which we live. Being the first commercial success in any endeavor is usually considered to be an enormous leap forward.

The ground-level construction of the open hearth furnaces at Homestead was arranged with the idea of facilitating both construction and operation of the furnaces, but the low-level placement of the furnaces in combination with the central casting pit was a return to some of the most nightmarish concepts and consequences of the English Bessemer plant and casting pit layout.[94] What led the Carnegie Company to abandon Holley's proven advantages of the high furnace and shallow pit? Did the fact that Homestead's Bessemer plant, which had already followed the English design, yet still became a success under Carnegie management, play a part? One can only surmise that this decision may have been influenced by their announced intent to cast large in-

Table 6.1. Open Hearth Steelworks in Pittsburgh (Allegheny County), 1886

Company	Number of Furnaces	Capacity (Gross tons per furnace)
Black Diamond Steel Works (Park, Brother & Co.)*	Not given	Not Given
Fort Pitt Foundry (Mackintosh, Hemphill & Co., Ltd.)	2	12
Homestead Steel Works (Carnegie, Phipps & Co., Ltd.)	4	35
Hussey, Howe & Co., Ltd.	1	35
Juniata Iron and Steel Works (Shoenberger & Co.)	2	12
Labelle Steel Works (Smith Bros. & Co.)	1	15
Linden Steel Co., Ltd.	2	10, 15
Millvale Rolling Mills (Graff, Bennett & Co.)	2	15
National Tube Works Company	1	10
Pittsburgh Steel Works (Anderson, DuPuy & Co.)	1	20
Singer, Nimick & Co. Ltd.	1	10
Soho Iron Mills (Moorhead & Co.)	2	15
Spang Steel and Iron Company Limited	3	10

*Park routinely refused to disclose information about his company, but the fact that he made open hearth steel is inferred first by his plant being included on this list from the *American Iron and Steel Association Directory*, as well as from other sources (*Recent Naval Progress*, June 1887, Navy Department, Bureau of Navigation, Office of Naval Intelligence, Government Printing Office, Washington, 1887, 242. It was also reported in an 1890 report from England, where it was stated that there were five furnaces at Park's works, ranging in size from twenty to thirty-five tons (*Journal of the Iron and Steel Institute in America, Special Volume of Proceedings*, E. & F. N. Spon, London, 1890).

Source: Directory to the Iron and Steel Works of the United States, The American Iron and Steel Association, Philadelphia, 1886, 115.

gots of sixty to one hundred tons or more, from multiple furnace heats, made in a single mould. Their plan was to roll plates weighing fifty tons.[95] The Park Brothers' plant was built at ground level to make it easier and cheaper to build. The Black Diamond Works' and the Homestead Works' furnace plans look remarkably similar, differing primarily in the arrangement of the tapping and teeming pits in relation to the pit crane. Henry Aiken worked at Park Brothers, but later he was employed at the Homestead Works to put the furnaces into operation. Both he and Julian Kennedy laid out the preliminary plan for this first shop.[96] With low-level furnaces, the charging doors and tap hole were easily accessible from the ground. A ladle pit was located in convenient proximity to the furnace, with a crane interposed between the ladle and casting (teeming) pit. These were some of the benefits highlighted by Horace Lash, Henry Aiken, and David A. Park (brother of James Jr.) of Park Brothers (Black Diamond Works)[97] in their patent application for the furnace design (number 365,936) filed on September 14, 1885. Their patent filing was months in advance of Carnegie's April 1886 disclosure about Homestead's expansion into open hearth steelmaking.[98]

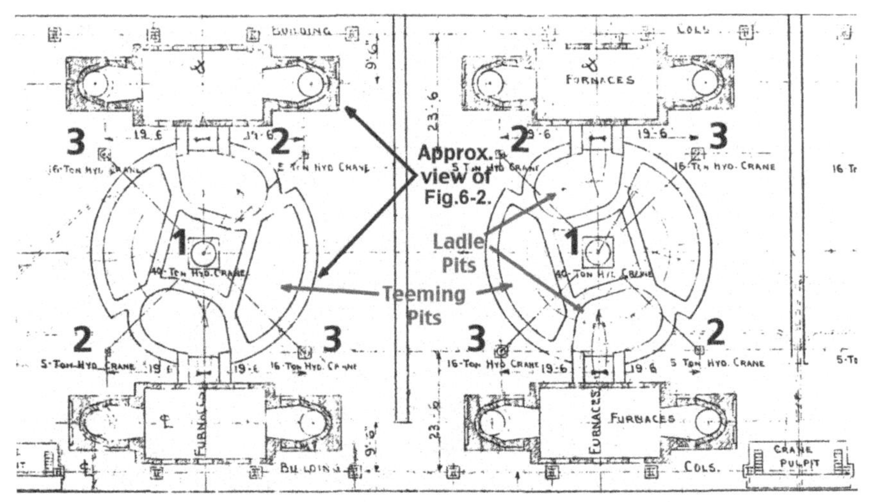

Figure 6.4. This enlargement of figure 6.3 is a view of the circular shaped central ladle and teeming pits of the Number 1 Open Hearth plant at Homestead. Each pair of furnaces had 2 ladle pits (one for each furnace) to receive molten metal from the open hearth and 2 teeming pits for teeming (casting) ingots. There were 5 hydraulic cranes to service these pits: one 40-ton ladle crane (marked #1) was used for both furnaces, in addition there was a 5 ton auxiliary crane (#2) for each furnace and one 16 ton ingot crane (#3) for each teeming pit. Each furnace initially had a 15-ton capacity. The view arrows give an idea of the approximate field of view of the photo in figure 6.2.
Archives Service Center, University of Pittsburgh, William J. Gaughan collection.

The new plant could easily have been designed with the two-level configuration typical of many other open hearth shops, such as Otis, where the charging floor was at a raised elevation and the ladles were at ground level, but that would have required a much taller building to house them, along with taller hydraulic cranes, because they were still top-supported at the roof trusses.[99] Some of the concern with using the two-level plan involved apprehension about the labor required to raise all the raw materials to the elevated level. The details of material handling on the high-level cupolas at the Bessemer shop at ET seem to have been implemented rather well there.[100]

The depressed central pit at the Homestead Open Hearth Shop was yet another matter. It had to be relatively deep, allowing for construction of the furnace proper with the tapping spout only about one foot above the shop floor. According to an anonymous worker (L. H.) from the plant:

> It is frightfully hot in the pit. A man can only stay there an instant. One man jumps down, strikes a couple of blows, and is pulled out again; another takes his place, and he in turn is succeeded by another. All are sweating profusely, and with the hot sand on the feet, and the fine, hot dust in their eyes, are miserable.

> Yet nobody falters or hangs back; each one jumps in, in his turn, and is hauled out laughing but exhausted. Think of getting down into a pit where the "heat" which has just been poured is standing on each side of you, while behind and in front are the walls so hot that they would char wood were it put against them![101]

The men were returned to harm's way again when they were frequently exposed to all the sparks, flames, fumes, heat, metal splashes, and spills that were credible dangers when tapping and teeming activity occurred simultaneously in a confined area. Only this time, the peril lasted longer because the fifteen, twenty, or more tons of molten steel made each time had to be run out through the smaller tap hole of the furnace instead of a mere five or ten tons that was poured quickly from the wide-open mouth of a converter. Additionally, because of the methods they used to tap the furnace into a ladle, slag would float on top of the steel and finally overflow the lip of the vessel and spill onto the floor of the furnace pit because there was not enough room in the ladle for both the steel and the slag. Men would have to go into the pit to face the formerly molten and now jelling slag and break it up into small-enough chunks so it could either be shoveled or loaded into boxes and moved away for disposal. This was another distasteful, hot job performed in a confined area, sometimes when the furnace directly above was full of "cooking" steel. All in all, the inherent problems with the return to an old English-style layout adapted from using converters were compounded by many of the idiosyncrasies of operation, introduced through the adoption of the open hearth process.

The #1 shop at Homestead was quickly expanded from the original four furnaces into one eventually encompassing eight. Shortly afterward in 1890 a new (#2) shop was built nearby to the south. This facility contained eight more furnaces. The quote found in the *McClure's Magazine* article about the pit conditions cited above was actually made about experiences and physical conditions found at Homestead's #2 shop, not at #1's. If the evaluation of the environment found in the teeming pit of #2 was so bad, one can only imagine how horrible it must have been at the original shop, since the layout of the newer facility was physically modified to address problems found in the arrangement of the #1 shop. As can be seen from the plant's actual floor plan given in figure 6.6 and the schematic representation found in figure 6.7, the mistake of the centrally located hydraulic crane with the ladle and casting pits positioned between pairs of furnaces was addressed at the #2 shop by adding two sets of rails between the two rows of furnaces. A full ladle of molten steel could be moved from the depressed ladle pit at each furnace and placed on a rail car, which was then pushed by a locomotive to the far end of the shop, where a pair of casting pits (mentioned above) were located and the teeming operation could then commence. (This picture, as rendered by employee

L. H., reflected the state of affairs found in the pit at this location.) Note the similarity in the layout of both of the Homestead shops when compared to the schematic of a Lash furnace at the Park Brothers & Company's Black Diamond Steel Works in Pittsburgh found in figure 6.8 that predated those at Carnegie's Homestead plant.

At the time that Carnegie's first two open hearth shops began operating, the charging of all raw materials into the furnace was done by hand. Furnaces of this early period had two, but usually three, doors on the charging floor (front) side. Some furnaces were built with access doors on the tapping side as well. In fact, when viewed from the charging floor, if it weren't for the extra brick structure for the gas and air ports at either end, an open hearth looked like it was all doors. The doors provided access so that the raw materials could be evenly distributed about the shallow hearth. That access allowed repairs to the refractory bottom and walls to be made more easily. In the beginning, the cold metallic charge of pig iron and scrap was most often added to the furnace, using a long bar with a handle (about fifteen or so feet in length) that had been flattened into a wide plate at the opposite end. This device was called the *peel*. Laborers would lift heavy chunks of scrap and pig onto the plate of the peel and, once loaded, a man or men on the peel would push it into the furnace and dump the metal in an appropriate place to evenly distribute the charge throughout the hearth (see figure 6.5). It took about six or eight men to charge a twenty-five-ton furnace. An estimate made by Reitell indicated that about forty unskilled laborers were required to perform this task in a shop of six twenty-five-ton furnaces.[102] Here is a comment from a former Bethlehem Steel superintendent: "This work was the hardest and most strenuous of the whole steel industry."[103] To feed a twenty-five-ton furnace with a crew of eight men, each man had to lift, carry, and place about three tons of metal. The Homestead furnaces were rated at thirty or thirty-five tons, equating to about four tons per man. Charging usually took less than thirty minutes. In a paper presented by Alfred Hunt to the American Institute of Mining Engineers (AIME), he indicated that charging twenty-five tons into a Lash-designed furnace at Pittsburgh averaged twenty-four minutes, using twelve men. That number included workers from the furnace crew, as well as the charging crew. The charging was accomplished using five doors (both front and rear) on the furnace. The stated time was based on charging data accumulated from five hundred consecutive heats.[104] Still, the task of rapidly handling three, four, or more tons of heavy scrap and/or iron was not the worst part of the ordeal for the men. Again, the matter was described in *McClure's*:

> Of the heat at a furnace door one can have no conception unless he has done the actual work. Four feet away the heat is sufficient to make your clothes

Figure 6.5. An early view showing a charging crew for a hand charged open hearth furnace. The men are congregated around the long handle of a peel. They are preparing to push a large piece of scrap that can be seen resting on the end of the peel opposite the men. Once the furnace door is raised, the scrap will be slid into the furnace with the peel. Members of charging crews were amongst the lowest paid employees in a steel mill and had to perform what was possibly the most difficult job in the mill.
Reproduced from: The Open Hearth, published by the Wellman-Seaver-Morgan Co., 1920.

> smoke; and there are times when a man must work directly up at the door. The skin contracts and seems about to burst; a steam rises from it, and the salty perspiration gets into the eyes and pours in streams from the face. After such work a man can sometimes actually wring water from his clothes and even pour it from his shoes.[105]

Here is another contemporary appraisal of conditions, and more to the point: "It was working aside of hell ahead of time."[106] Men would often work until they were physically spent, especially in hot weather or when the circumstances demanded, such as when there were a series of hot, strenuous jobs that followed one after another, for instance, when a number of furnaces were charged and/or tapped in quick succession. It's a strange feeling when a body's bones and muscles no longer respond to the brain's commands. Sleeping on the job (in the form of a quick nap) was unofficially condoned. This was man-killing work, and these men could not be expected to do it continuously for twelve or more hours without a rest, especially at night or when a man had to work the *long turn*. The long turn consisted of a prescheduled shift of twenty-four consecutive hours of work, allowing a man to switch from working days to working nights. This occurred every other week: "One who has fallen asleep in spite of his efforts to keep awake . . . sleeps the sleep of utter exhaustion."[107]

Completed in the fall of 1890, the new #2 shop consisted of two rows of four furnaces, each with a capacity of twenty tons (twenty-five tons as reported elsewhere). Apparently, some of the innovative steps in the area of material handling that were taken by the Carnegie Company in this furnace shop went far beyond the movement of the molten steel ladles away from the furnaces to a casting pit at the far end of the building, where ingots as large as sixty tons could be cast. In an important but not well-documented move introduced at the Homestead Works was the use of mechanized charging machines designed by Kennedy and Aiken in an attempt to replace the extensive need for men.[108] Apparently, this action was taken in order to eliminate men where they were most exposed to the detrimental effects of the environment in close proximity to the furnace doors. It is possible that they started mechanizing the operation as early as 1889. There is evidence of this in a reference to the rate of pay for a charging machine operator that was listed for a (sliding scale) contract that included that particular year at the Homestead open hearth. Furthermore, they were using charging boxes about that time, as per an article in *The Iron and Steel Institute in America in 1890 (Special Volume of the Proceedings)*. Since no more references could be found, the date is uncertain.[109] This is not to suggest that they mechanized out of a deep regard for the men. These jobs were so physically demanding that the work was nearly impossible to accomplish without an extremely large but low-paid crew that quickly became depleted by the brutal conditions. There were two new machines, one for each row of four furnaces. Each one had to be placed at the appropriate furnace using a locomotive crane. Bringing scrap and other material to the furnace on cars in small boxes to be charged by machine was also introduced. The locomotive crane also might have been used to handle the boxes of material (see figure 6.12). Whether this was successful is not known.[110]

In 1887 Samuel T. Wellman is believed to have developed the first documented equipment (at least something that was relatively successful) used for the mechanical charging of open hearth furnaces at the Otis Steel Company in Cleveland, Ohio. His hydraulically actuated device, while functionally acceptable, proved to be inadequate due to a number of hydraulics-related problems. In 1890 Wellman's early floor design called for the use of steam for power.[111] The breakthrough occurred with his application of Westinghouse electric motors to power all of the functions, except one, on a floor-mounted machine built for a company near Philadelphia

(Thurlow) in 1894.[112] The machines were very simple yet rugged. They had a long peel with an end that could fit into a pocket molded into one end of the charging boxes. A locking pin would then be engaged (the only manual function), the peel would lift the box from the car, move it forward through an open door of the furnace, and rotate it to dump the load of scrap or pig iron, iron ore, and limestone. The empty box would be returned to the car and another full box would be retrieved, about one ton at a time, until the furnace was filled. There were usually three and sometimes four charging boxes per car.[113] Because of Wellman's success with this new equipment, sometime before or during 1895 Carnegie bought and began using at the #2 open hearth some specially modified electric charging machines that were manufactured by Wellman.[114] (See figure 6.13.) His equipment also used small boxes (roughly six feet by two feet by one-and-a-half feet)[115] that contained about one ton of scrap or pig.[116]

Use of Wellman's modified machine at Homestead, which was capable of charging at a rate of two tons per minute,[117] revolutionized the way in which material was handled and charged at large-volume open hearth steel shops in this country and around the world. By 1897, the Homestead works had six Wellman charging machines in operation,[118] indicating the degree of their success, when compared to Carnegie's original mechanical charging experiment circa 1890. A 1901 sales catalog from the manufacturer showed that other large companies such as Cambria, Pennsylvania Steel, Pencoyd, and Phoenix (all located in Pennsylvania) and several others were also using Wellman's machinery. Some had as many as three machines each, while Illinois Steel in Chicago was listed as having only one. That same publication indicated that by this time Carnegie employed fifteen of the devices for his open hearths, twelve at Homestead and three in the new shop at Duquesne![119] The British claimed a two-and-one-half hour savings by using a machine instead of hand charging a forty-ton furnace that usually took about three-and-one-half hours to charge.[120] Citing information presented by Hunt, only twenty-five to thirty minutes was the customary time needed to hand charge a twenty- or twenty-five-ton furnace in the United States. (Extrapolating from that same data furnishes a rough equivalent of one hour for a forty-ton furnace.) Therefore, even if the times given in America were understated to appear more favorable, the time saved does not appear to have been the primary justification for automating the process in this country. This is supported by the fact that factories with only one or two furnaces owned and operated Wellman or other makes of mechanical charging machines. These new Wellman machines ran on tracks mounted in the floor of the shop as opposed to the original design, which had the machine suspended overhead on rails located above the top level of the furnace.[121]

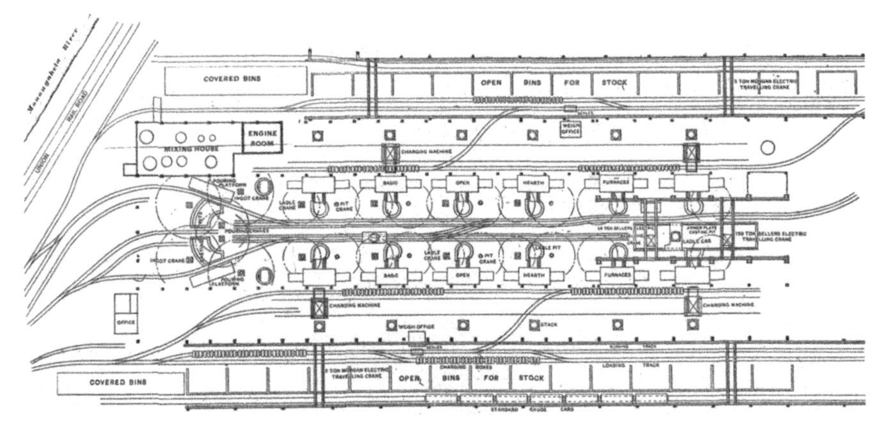

Figure 6.6. Plan layout of the #2 Open Hearth Shop at Homestead. Note the two sets of tracks between the two lines of furnaces which led to a pair of semicircular casting pits at the left of the drawing. This permitted the ladles of newly tapped molten steel to be transferred to these pits by locomotives and then be teemed there into ingots, eliminating some of the problems alluded to by anonymous author LW in the McClure's Magazine article with regard to the central pits at the #1 Shop. Later, shown at the right end of the drawing, a rectangular casting pit was built for teeming large ingots (see figure 6.14).
Iron Age, Vol. 60, 1897.

Another important but not well-documented advancement that was introduced at about the same time involved the use of electric overhead traveling (EOT) cranes at Homestead. Photos from a Carnegie Museum of Art collection in Pittsburgh circa 1893 to 1895 verify that electrically driven pit cranes were already in operation at the open hearth shop for use in casting large armor plate ingots as well as numerous other applications.[122] An article from the *Metallurgical Society* concerning the history of steelmaking indicated that electric cranes were installed prior to 1891 at a number of steel plants in the United States, including Homestead. A U.S. Navy report placed an overhead traveling crane (probably not electric) in the plant as early as 1889.[123] The first crane for which a reference could be found that was designated as a pit ladle crane in an open hearth plant was a sixty-ton crane built for the Homestead Works in 1897. Photos from the Carnegie Museum clearly show that a sixty-nine-ton crane at the armor casting pit in the #2 open hearth shop was in operation circa 1893.[124] (See figures 6.14 and 6.15.) Overhead traveling cranes existed before electric motors but were rope driven, belt or shaft driven, steam driven, or even powered by hand. These cranes had limited capabilities and were very complicated to operate. The rush to adopt modern equipment made an overhead crane seem an obvious choice, especially when the utility of the equipment was expanded because their operation wasn't lim-

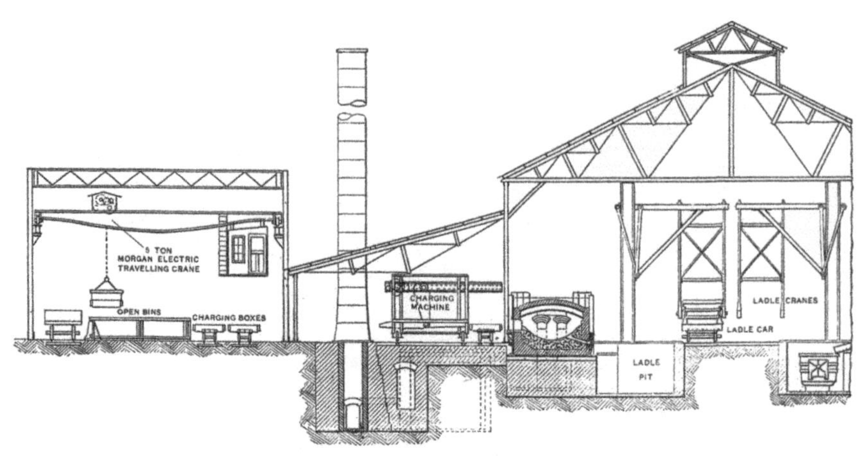

Figure 6.7. This sectional view shows the change made from #1 Shop to the central isle of the second plant. The furnaces on the right have been cropped from the diagram. The pairs of hydraulic ladle cranes for loading them from the casting pit onto cars at each furnace are indicated. The addition of the Wellman electric charging machines, which postdate the 1890 construction of the plant, are also shown.
Iron Age, Vol. 60, 1897.

ited to a confined area of the factory floor, as was the case when comparing them with fixed hydraulic cranes. Still, hydraulic floor cranes can be seen in operation in photos of a Bessemer facility as late as the 1940s (see LTV Collection of the Western Reserve Historical Society[125]), lending credence to the notion that not everyone in the industry got that same message.

Coincident with the introduction of electric-powered traveling cranes and the new charging machines, Wellman invented a large electric magnet in 1889 that was originally devised for lifting billets.[126] (It was patented in 1896, but I believe it was later rejected.) Besides making it easier to handle billets and large plates, this magnet was a perfect match for loading scrap metal and pig iron into the charging boxes, which then quickly filled the furnace. Thus, there were three complementary time- and labor-saving inventions introduced in relatively quick succession that enhanced the operations of large-volume, basic open hearth steel shops during a critical time in their evolution. These inventions were largely related to the development of electric motor technology.

At the same time such advancements were being made in steelmaking in the Pittsburgh region, both direct and alternating current motors, together with power generation and distribution systems, and were being engineered and manufactured at the Westinghouse Electric factory located along Turtle Creek (a tributary of the Monongahela River at Braddock) in East Pittsburgh, just a relative stone's throw from Carnegie's three steel-producing plants

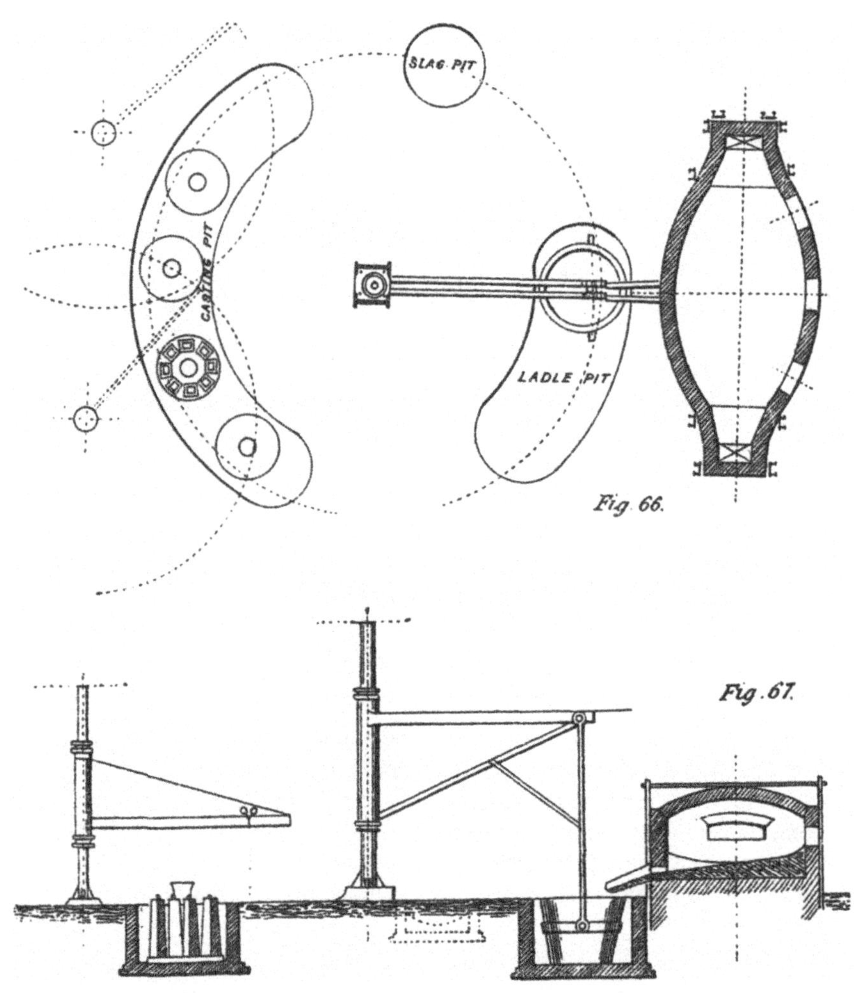

Figure 6.8. Plan layout for the Park Brothers' open hearth furnace at Black Diamond. The teeming or casting pit is opposite the furnace, not centrally located between rows of furnaces as with the Homestead plant, otherwise very similar. Also, the ingots were bottom poured as indicated by the "trumpet" or standpipe, central to the ring of ingot moulds.

Journal of Iron and Steel Institute, Special Edition, 1890.

in the valley. All of these new devices helped make possible the enormous strides taken to reduce the need for large numbers of low-wage hand laborers, traditionally employed in steelmaking operations. Further insight into the extensive use of electric motors at the Edgar Thomson Works, most instituted prior to the dawn of the twentieth century, can be found in an article titled "Electricity in Modern Steelmaking" by John Hayes Smith.[127]

There still remained many other mundane tasks requiring manual laborers to perform, such as:

- cleaning slag spilled in ladle pits after it overflowed the ladle when tapping each and every heat
- cleanup of spilled steel after ladle and furnace breakouts
- digging out the worn refractory in ladles and furnaces when they required repair, relining, or rebuilding
- handling the enormous volumes of materials shipped and received

Although not directly connected to the use of electric motors, electrical illumination had been installed in the Carnegie plants for a number of years. By 1889 the Homestead Works had electric lights.[128] The first use of electrical lighting in the American steel industry was typically in the form of an electrical arc, reportedly installed at the Edgar Thomson Works in 1881.[129] Arc light was very harsh. Gaslights were also used. As an aside, it is interesting that by 1889 Homestead had installed telephones to connect important operations so reliance on messengers could be eliminated.[130]

> Calendar of one day from the life of a [anonymous] Carnegie steel workman at Homestead on the open hearth, [working as] common labor:
>
> 5:30 to 12 (midnight)—six and one-half hours of shoveling, throwing and carrying bricks and cinder out of bottom of old furnace. Very hot.
> 12:30—Back to the shovel and cinder, within few feet of pneumatic shovel drilling slag, for three and one-half hours.
> 4 o'clock—Sleeping is pretty general, including boss.
> 5 o'clock—Everybody quits, sleeps, sings, swears, sighs for 6 o'clock.
> 6 o'clock—Start home.
> 6:45 o'clock—Bathed, breakfast.
> 7:45 o'clock—Asleep.
> 4 P.M.—Wake up, put on dirty clothes, go to boarding house, eat supper, get pack of lunch.
> 5:30 P.M.—report for work.[131]

A third open hearth shop, initially consisting of sixteen forty-five-ton furnaces but soon expanded to twenty-four fifty-ton furnaces, was built at Homestead beginning in 1897. Similar to the first two shops, #3 was constructed with a low-level (at grade) furnace layout. However, Wellman electric charging machines as well as electric overhead traveling cranes in the pits, on the charging floors, and in the stockyards were used from its very beginning (see plan in figures 6.9 and 6.10). Ingots were teemed in moulds on cars, and there were hydraulic pushers at the pouring platforms arranged

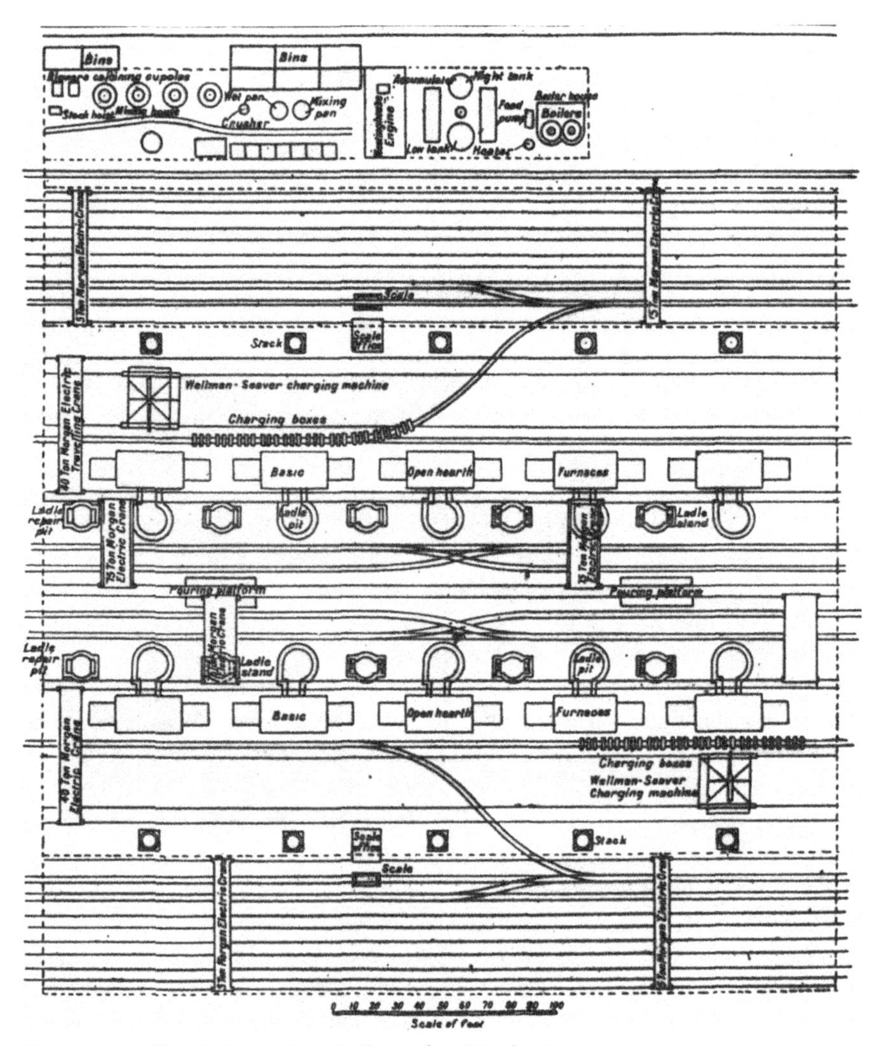

Figure 6.9. The #3 Open Hearth Shop of 1897, plan layout.
Iron Age, Vol. 61, 1898.

to move the moulds with the cars into position under the ladle. The seventy-five-ton pit cranes were large for the time. There were also forty-ton cranes to charge the molten iron into the furnace. The remainder of the materials—the scrap, pig, ore, and stone—was left to the Wellman charging machines. At the time, a fifty-ton heat could be comprised of 40 to 50 percent scrap, the remainder being hot metal, depending on the availability of both scrap and hot metal. At the beginning of operations, all of the hot metal would come

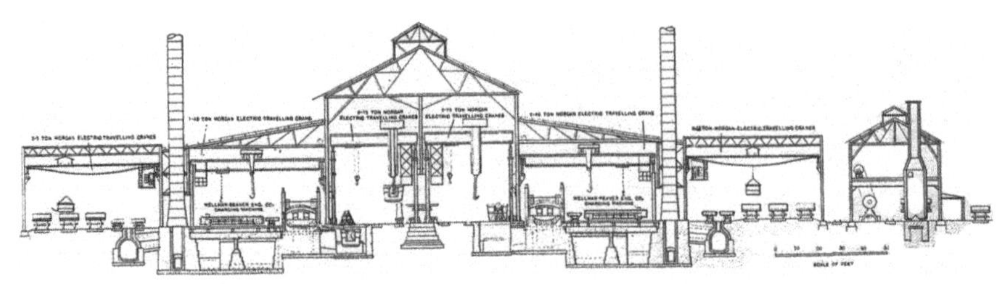

Figure 6.10. Cross section through #3 Open Hearth Shop showing differences in layout from the #2 Shop, including overhead traveling cranes for both the charging and the teeming floors, and low profile Wellman charging machines. The rapid evolution of plant design can be seen between the primitive layout of the #1 Shop with its hydraulic cranes and central teeming pits in figures 6.2 through 6.4 when compared to the more sophisticated plan above with its electric cranes and chargers and ability to cast ingots on cars.
Iron Age, Vol. 61, 1898.

from either Edgar Thomson or Duquesne, with the supplier being the driving force in the selection of the blend. An all-cold charge could be made as well with machine or sand cast pig replacing the molten iron, but the cycle time and the fuel rate for the furnace would rise considerably. Charging times could rise as well, because a Wellman charger filled the furnace at about two tons per minute.

Construction began on September 10, 1897, and the first heat was tapped on April 18, 1898, only seven months later, although the entire shop was not yet completed. A mixer for hot metal was added at one end of the shop in 1898. Carnegie experimented with a hot metal charge in the furnaces at Homestead as early as 1892, long before the third shop was ever contemplated, using cupola metal instead of molten iron brought directly from the blast furnace. Apparently, this was an attempt to increase productivity by decreasing the cycle time of the open hearth furnace itself.[132] There was neither a bridge to ET nor blast furnaces at Duquesne from which to procure hot metal until a short time before the #3 shop was built. When it was completed, this facility was the largest open hearth plant in the country.[133]

There is an extremely important correlation that complemented the introduction of basic open hearth steelmaking at Homestead. In fact, it had a profound impact on its establishment there, and this association is oddly enough directly related to the enormous rate of Bessemer steel production that had occurred in both the Edgar Thomson and Homestead plants. To my knowledge, this concept has never been properly or adequately scrutinized. It provides some evidence to help explain the specific reasons behind the development and rapid expansion of the open hearth production facilities at the Homestead Works.

Once, many years ago, there was an article (cited here without reference) that can be paraphrased thusly. It described a Bessemer steel plant as a factory that appeared to be surrounded by a huge junkyard. Scrap was stored everywhere.[134] Scrap was not typically a raw material in the Bessemer process; it was used only in the few instances where it was required in small quantities to act as a coolant in order to reduce overly high metal temperatures. If the scrap wasn't routinely used, why was it there? There must be something more to comprehend from the mental image of a junk-riddled plant. High converter temperatures usually occurred when the molten iron contained above-normal levels of silicon. This would often lead to overheating of the metal bath, because silicon burning was the primary source of heat in a Bessemer reaction and it took longer to burn out the higher levels. Even though that particular reaction occurred during the early part of the cycle, it still added to the blowing time. Overheating could also occur with normal levels of silicon in the iron, following a series of quick, consecutive heats, when the converter vessel had not been permitted to cool. Recall that Bill Jones had devised a method of lowering the temperature of the molten metal by injecting steam or water into the blowing air, avoiding the delays involved with "scrapping" or adding cooling scrap to the converter. By saving the time required to perform the scrapping function, his procedure thereby helped to maintain high production levels. This seems to indicate that time was the far more precious commodity rather than any monetary value gained through using the scrap that otherwise would have been consumed.

There was an associated value assigned to the scrap accumulated at the Edgar Thomson Works, and it was enormous. According to his 1876 annual report to Carnegie for ET's first full year of operation, Shinn reported that the scrap was worth $102,806.25, or slightly more than 5 percent of the total assets ($1,914,496.57) of the company. (See figure 6.11.) Whether that value is actual or an estimate is not certain.[135] Unlike today, many worn wrought iron and steel scrap items (especially rail) could be purchased on the open market by small firms that had a very low overhead for rerolling or repurposing, so there was some identifiable value that could be assigned. However, used steel required a more powerful mill to roll it than wrought iron, and much of the scrap generated at a Bessemer mill probably would have had too many imperfections to be acceptable for repurposing; problems such as:

- extensive piping in the crop ends
- too many blowholes
- irregular shapes as with skulls
- other types of contamination

Office of The Edgar Thomson Steel Company Limited

Pittsburgh January 26th 1877

To the members of The Edgar Thomson Steel Company Limited

The undersigned submits this, his first Annual Report, upon the financial condition of your Association, on December 31st 1876, and of the operation thereof to that date.

The Balance Sheet of the Association on December 31st 1876 shows as follows

Assets.

Cost of Works	$1.096.432.84	
Real Estate Cost	259625.93	
Tenements	33.072.40	
Discount on Bonds sold,	12100.00	
Total Capital invested		$1 401 231.17
Stock on hand.		
Manufactured Product	$29.641.62	
Materials for use	117379.96	
Scrap	102,806.25	
		249.827.83
Available Assets		
Cash	$26.810.92	
Bills Receivable	64828.45	
Book Accounts	171.738.20	
		263.377.57
Total Assets		$1.914.496.57

Figure 6.11. The 1876 Annual Report for the Edgar Thomson Steel Works prepared by Shinn (first full year of operation). Reproduced here from *The Inside History of the Carnegie Steel Company*, James Howard Bridge, 1903, page 98. According to the line for scrap ($102,806.25) verses the final line stating the total assets ($1,914,496.57), scrap represented a value equivalent to about 5% of all the assets of the company.

Figure 6.12. Photo of charging boxes on cars in front of either the #1 or the #2 open hearth shop at Homestead circa 1893–1895. The boxes are possibly of the early Carnegie Company design as the circular sockets at the charging machine end (near) of the semi-cylindrical boxes are different from the square socket of the Wellman design in figure 6.13.
Reproduced with permission, BLH Dabbs photo, Carnegie Museum of Art, Pittsburgh, PA.

Figure 6.13. A very early 20th Century photo of a high profile Wellman charging machine designed for close quarters at the Homestead Works of the Carnegie Steel Company. Photo is likely the Number 2 open hearth shop c.1908, as per the schematic diagram of that shop in figure 6.7.
Library of Congress collection.

Figure 6.14. Large 69 ton capacity Morgan electric overhead traveling pit crane at the Number 2 open hearth shop at Homestead circa 1893–1895. The view shows the process of a continuous pour of multiple open hearth heats for large forging ingots, likely for armor plate. A ladle from a more recent heat (higher) is being teemed into a ladle on a carriage from an earlier heat (lower), which flowed into a large ingot mould in the pit at the bottom through a standpipe or "trumpet" into the bottom of the mould. This permitted an unbroken flow of steel into the mould, which was larger in capacity than a single furnace was capable of filling.

Reproduced with permission, BLH Dabbs photo, Carnegie Museum of Art, Pittsburgh, PA.

Figure 6.15. The Carnegie Steel Company had many electric overhead traveling cranes in operation early in the history of these devices. This view circa 1893–1895 shows a number of them in operation at the structural steel storage yard at Homestead.
Reproduced with permission, BLH Dabbs photo, Carnegie Museum of Art, Pittsburgh, PA.

Acid open hearth shops of the time were typically small, low-production facilities. Open hearth furnaces were often associated with crucible steel and puddling furnace operations as well as foundries, but the previously mentioned Otis Steel Company in Cleveland was the first shop built in the United States solely for the production of open hearth steel (with two seven-ton acid furnaces). It went into operation in 1874, but a new shop with four fifteen-ton raised furnaces (also acid) was operating by 1881, more than five years ahead of Homestead.[136]

Basic open hearth shops could consume most phosphoric scrap like wrought iron, while acid shops (both Bessemer and open hearth) could not. Unlike the special conditions required by a Bessemer facility, acid and basic open hearth shops could readily use "home scrap"—the scrap that was generated during the steelmaking process itself by its own plant. The dollar value of the scrap differed depending on whether it was created at a Bessemer shop or at an acid or a basic open hearth shop. H. H. Campbell asserted that Bessemer scrap had a lower value relative to that of pig iron at a plant's own Bessemer facility, whereas for an open hearth, that same scrap (home scrap) had more value than pig iron. This was due to the fact that in order to intentionally use Bessemer scrap in a converter, you would also have had to use higher silicon iron to generate sufficient heat to melt the scrap. That would require more time to blow as well as the extra time needed to add and melt the scrap.[137] These points verify the value of Jones's practice of adding steam to the blowing air. In a basic open hearth, using either Bessemer or home scrap presented none of these problems. Pig iron could be replaced and generally cost more than scrap (as the in-house scrap had value but no monetary cost). So scrap ultimately made a financial contribution to profit through reducing the final cost of producing the end product.

Scrap also proved its increasing value in that its use was required in the open hearth as a raw material when adopting the pig and scrap (Siemens-Martin) process. This was not true for the pig and ore (Siemens) furnaces or in Bessemer practice, where scrap was not a prerequisite. In this early era there was yet another benefit from a basic furnace being capable of consuming contaminated scrap containing certain levels of phosphorus. Due to its ability to use readily available scrap, the Siemens-Martin process was the most widely adopted open hearth practice in the United States.

The Edgar Thomson Works had been rolling rails almost exclusively for over a decade before the adoption of basic steelmaking at Homestead. In Homestead they had been making Bessemer rails as well as structural steel for about five years prior to the adoption of the open hearth. Steel rails were replacing the previously laid wrought iron track at a phenomenal rate. Some of the old rail could be relaid on secondary lines (in yards, sidings, and such), some could be rerolled, but most of the available iron rail could only be converted to steel in a basic lined furnace (since it was phosphoric). Consequently, at the onset that meant it could only be done efficiently at Homestead. Also, the end products from these Homestead furnaces were used to produce higher-quality goods such as plates and structural steels that commanded a higher selling price than rails. In the early 1890s, armor plate could be sold for as much as $850 or more per ton, while rails were only fetching in the $20 to $30 range.[138] However, there were many additional costs associated with producing armor above and beyond the price of scrap; for example, alloying and accounting for the extra operational and capital costs involved with forging and case hardening by a system known as the Harvey Process. Then, too, the market for this product was extremely limited.

At facilities like ET and Homestead, which held world records for output and were in the top rank of the world's Bessemer steel producers, scrap production must have been enormous. The use of scrap that was generated in house in huge quantities and available from within the company at no additional cost would have given Carnegie an enormous competitive advantage at the Homestead open hearth plant, because all of the metallic charges into the furnaces at that time were made cold and a high percentage of scrap to pig could have been used. After all, Carnegie was producing in the range of 30 to 40 percent of the open hearth steel made in the United States in what was the largest such facility in the country at that time.[139] So the functional importance of the early use of scrap in this way apparently was an important factor, and while only one factor, it was one that heretofore has never been addressed. Today, the reporting of scrap generation is a significant tool for determining the efficiency of each step in the steelmaking process. Yet with respect to the Carnegie era, finding such data is a challenging task at best.

In an effort to discern the approximate volume of scrap produced by the Edgar Thomson Plant, it is necessary to be creative in using the limited information available. A good estimate can be culled from reports in various journals that state both the ingot and rail production (and billets as well, if sold in that format) at the plant. Scrap would simply be the difference between the two. That information is shown in table 6.2 below, and its reliability is better established by stating it as a percentage of ingot production.

Based on the information from table 6.2 below, it appears that a reasonable estimate of the percent of scrap generated at the ET Works during the time span represented was approximately 18 percent. The first two years of data indicate a higher 20 percent to 21 percent range, but an advancement in technology introduced by Jones about 1878 that used steam to compress the molten steel while in the ingot mould resulted in a reported yield savings of 2.6 percent over previous years. So, obviously, they were concerned with addressing the causes of this loss and kept close track of that data.[140] The 18 percent average, together with the later estimates from the table, support that

Table 6.2. Estimate of Scrap Generation and the Edgar Thomson Works between 1876 and 1889

Time Frame	*Ingots Made (tons)*	*Rails Made (tons)*	*Scrap Made Col. 2 Minus 3 (tons)*	*Scrap Made as a % of Ingots*
1876[a]	42,112	33,258*	8,854	21.0
1877[a]	65,427	52,290*	13,137	20.1
April 1880[b]	3,433	2,823	610	17.8
Nov. 1880[c]	13,166	11,109*	2,057	15.6
1880[f]	123,303	100,094	23,209[g]	18.8
1886 (1 day, est.)[d]	900	700	200	22.0
March 1889[e]	31,120	25,342	5,778	18.6

a—*Engineering* XXV, London, 1878, 382. Represents one complete year of output.
b—*Iron and Steel Institute*, no. 1, 1881, London, 274.
c—Bulletin, *American Iron and Steel Association*, December 15, 1880, Philadelphia, 308.
d—"Pittsburgh's Progress," *Industries & Resources*, Thurston, 1886, 100.
e—*Franklin Institute* XCVII, 1889, Philadelphia, 366.
f—*Iron and Steel Institute*, no. 1, 1881, London, 140.
g—This does not account for 4,262 tons of merchant steel sold (129), making scrap more like 15.4 percent. However, there was 5,862 tons of scrap listed as rolling losses (blooming and rail mill) that were not counted, bringing the value back to 18.8 percent.

*Includes billet production.

claim to some extent. It is not certain how long Jones's technology was used, but lower scrap production in later years could also have been the result of larger ingots and longer finished rails, cut to length by hot saws, resulting in fewer crop ends. The 22 percent figure from 1886 was based on estimated production for only one day and is not significant. A credible estimate of the actual tons of scrap produced can be determined by applying the 18 percent figure to the data for rail production found in table 6.3 below. This covers the years from the beginning of Edgar Thomson to the commencement of the second basic open hearth shop at Homestead.

The total volume of scrap produced during the time frame covered by table 6.3 below is only a rough calculation. While the approximate 476,976 tons of steel scrap generated had only an extremely limited utility at the Bessemer shop, that same material had an enormous value in the open hearth shop at Homestead. This figure does not even consider the scrap generated when billets were made directly for sale, nor any product that was produced at the Homestead Bessemer shop during this same time span, and it does not count scrap from spills and skulls. So taking that into account, it can be said with confidence that through constructing the basic shops at Homestead, well over one-half-million tons of a valuable commodity was available for Carnegie to use for no additional cost, and, more importantly, it was transformed into

Table 6.3. Estimated Scrap Produced at ET Works as a Function of Annual Rail Production (based on 18 percent average scrap generation per ton of ingots from Table 6.2)

Year	*ET Rail Production (Gross tons)*[a]	*Estimated Scrap Produced*[b]
1875	5,853	1,285
1876	32,226	7,074
1877	48,826	10,718
1878	64,505	14,160
1879	76,044	16,693
1880	100,095	21,972
1881	151,507	33,258
1882	143,561	31,513
1883	154,892	34,001
1884	144,090	31,630
1885	126,656	27,803
1886	173,001	37,976
1887	192,999	42,366
1888	148,293	32,552
1889	277,401	60,893
1890	332,942	73,085
Total estimated scrap produced 1875 to 1890, inclusive		**476,976**

a—From table 8.1.
b—Not accounting for any nonrail billets produced.

more profitable products such as beams, plates, and armor. The open hearth process itself was a huge source of this self-generated raw material: "For every ton of ingots produced and rolled, 25 to 30 percent by weight will be returned to the open-hearth stockyard as mill or home scrap for remelting [this is based on modern estimates]."[141]

Andrew Carnegie stated in testimony before the Committee of Naval Affairs that it took three tons of open hearth steel to make one ton of armor plate. In other words, two-thirds of a 150-ton ingot (one-third each from the top and bottom) had to be removed as scrap in order to reach the most sound section of steel to forge into armor.[142] The manufacturing of armor plate was probably the most wasteful means of producing an end product. Carnegie had two or sometimes three (circular) furnaces with removable roofs operating (two of them are clearly seen at the left in the plan view in figure 6.3) to use the large pieces of scrap that couldn't fit through a normal-sized open hearth furnace's doors by charging them through the top of the furnace with a crane; parts like skulls, large parts of armor plate, ingot butts, spills from leaking ladles, furnaces breakouts, or accidentally discharging into the pit, and so on.[143] While the amount of waste generated from making armor plate was huge, yearly production of steel for armor was not. Only several thousand tons were made. So it was not as significant as it might seem. Previously, large chunks of scrap like skulls and spills (prior to the advent of acetylene cutting torches and "skull breakers") were frequently found to be unusable because of their size, and they were either unable to break them into a usable size or it was impractical to do. They were simply taken to some out-of-the-way location in the plant and abandoned or buried.[144] Production of armor plate seemed to be a source of some pride for Carnegie, although it brought with it much tribulation when a controversy arose about the quality of some plates. That tribulation however, was compensated for by large profits.

"In theory the Bessemer should produce a perfect steel." However, if you overblow it by even a few seconds, you begin to oxidize, or "burn," the iron. Of course, if you seriously overblow the heat you make a heap of useless junk. Underblown steel was not good, either. Higher carbon could render it hard and brittle. Sulfur and phosphorus levels increase when making steel in acid vessels because those contaminants were not eliminated. At the same time, there is a loss of iron caused by blowing that in turn reduced the volume that contained these detrimental chemicals, thereby increasing their proportions in the heat. The high temperature at which this steel was teemed favored casting small ingots, but it was found to create serious problems with larger ones. Here is very brief summation of an 1889 analysis: the U.S. Naval Institute determined that steel produced in a Bessemer converter was unacceptable for use in making armor, even though the Bessemer process was found to be more economical than the open hearth.[145]

With basic open hearth furnaces, the Carnegie Company entered into the realm of making better-quality, higher-priced steels than their Bessemer operations. It refined its operations through the introduction of electrified heavy machinery, chargers, and overhead traveling cranes to remove many tedious jobs, and simultaneously to increase productivity and facility utilization. As these were basic furnaces, the unusable scrap being generated at their Bessemer plants, scrap having little or no utility, now became a valuable commodity at the open hearth shops. After experimenting with the use of cupola iron to speed up open hearth cycle time by eliminating melting time for this part of the cold charge, the use of hot metal, brought via rail cars to the mixers directly from the blast furnaces at Duquesne and Edgar Thomson, was adopted as soon as it became feasible. This motivated the purchase and expansion of the Carrie Furnaces, which were located almost directly across the river from the plant.

Table 6.4. American Steel Production (Tons of 2,240 Pounds)

Year	*Bessemer Ingots*	*Open Hearth Ingots*	*Comments*
1877	500,524	22,349	
1878	653,773	32,255	
1879	829,439	50,259	
1880	1,074,262	100,851	
1881	1,374,247	131,202	
1882	1,514,687	143,341	
1883	1,477,345	119,356	
1884	1,375,531	117,515	
1885	1,519,430	133,376	
1886	2,269,190	218,973	Homestead #1 open hearth opens late in year.
1887	2,936,033	322,069	
1888	2,511,161	314,318	First commercial basic open hearth steel, Homestead.
1889	2,930,204	374,543	
1890	3,688,871	513,232	Homestead #2 open hearth shop begins late in year.
1891	3,247,417	579,753	
1892	4,168,435	669,889	
1893	3,215,686	737,890	
1894	3,571,313	784,936	
1895	4,909,128	1,137,182	Wellman chargers installed, Homestead open hearth
1896	3,919,906	1,298,700	
1897	No data	1,608,671	Homestead #3 open hearth shop planned.

Source: William B. Phillips, *Iron Making in Alabama*, second edition (Montgomery: Alabama Geological Survey, 1898), 308–9. The numerical data was assembled from reports produced by James Swank, manager, American Iron and Steel Association.

Chapter Seven

Duquesne

In 1888 some of the same men who were associated with Andrew Kloman in establishing the failed Pittsburgh Bessemer Steel Company in Homestead joined together to form the Allegheny Bessemer Steel Company in Duquesne after Kloman died. A few of the principals involved were William G. and David E. Park from the Black Diamond Steel Company and E. L. Clark of the Solar Iron Works (not to be confused with William Clark, also from Solar, who was largely responsible for the poor management of PBSCO). Actually a reorganization of the incomplete Duquesne Steel Company of 1886, this facility was built with a two-vessel converter plant of seven tons capacity each, a blooming mill, and a rail mill. The Pittsburgh firm of Macintosh & Hemphill built the rolling train, which was set up to roll rail directly from the ingot phase straight through to the final product without the traditional reheating step that occurred between making a bloom and then rolling the rail. This is the same firm that built the rolling mills for Kloman in Homestead and, as in the past, the Duquesne mill was of the most advanced design. Now for a second time there was a local mill more technologically advanced than Carnegie's. In this instance all of the physical handling, as well as the time and effort involved in performing a complete reheating step, was saved. The new company suffered from a combination of poor management and erratic relations with the labor force. Carnegie's false accusation that the new direct rolling method caused defects in the rail did not help, either. Frick negotiated a deal and bought this plant in 1890 for the Carnegie Company without investing any capital. He paid for it solely with bonds.[1] Not only did they continue to use direct rolling at the "new" Duquesne Works, they also adopted the practice at Homestead and Braddock, too.[2] Nothing more was said about any flaws in the rails.

THE DUQUESNE REVOLUTION

> In the United States (which has taken the lead in this respect) the average yearly product of a blast furnace was, in 1874, 6,298 [tonnes]; in 1880, 9,369; and in 1895, 44,953 metric tons . . . In Germany the product (1895) per furnace was about 28,840 metric tons; in Great Britain (1895) 23,319 metric tons.[3]

In 1895 the Carnegie Steel Company embarked on a project to build a blast furnace plant at Duquesne that was capable of producing about 220,000 tons per year (tpy) per furnace,[4] or about five times the average output per furnace of the United States as reported above for 1895. The best furnace production that could be expected at this time was between three hundred and four hundred tons per day (about 146,000 tpy at 400 tpd). Occasional spurts of five hundred tons or slightly more per day were possible, but those rates were not sustainable.[5] There was simply no space on the top platform of the furnace for the loaders to move about while loading the bell and/or neither room nor cycle time on the elevators to hoist any more raw materials to the top.[6] Yet at Duquesne, notwithstanding the fantastic advancements that had been made in ironmaking at Edgar Thomson, the projected six hundred tons per day output per furnace was fully 50 percent more than the maximum that could be expected from the best furnaces in the world at that time. And what seems to be more striking, based on the information given above, this was roughly a 125 percent greater make of iron in a single day than the average British furnace was producing in an entire week. The new plant of four furnaces at one location achieved the equivalent output of thirty-five average blast furnaces in the United Kingdom. This is more impressive, considering that the Americans surpassed the British as the world's leading pig iron producer in 1890, only five years before!

It will probably not be surprising to know that the key to accomplishing this fantastic increase in output was mechanization. The scale of the new furnaces was not overly large, but they were well endowed with stoves, blowing engines, boilers, and more. However, this was not merely the automation of a small portion of the process but was a systematized upgrade and advancement of the entire blast furnace material-handling operation that was undertaken on an unprecedented scale. It was such an extraordinary achievement that once the transformation of the facility proved fruitful, and for decades afterward, steel industry experts in this nation and around the world would refer to it both in print, as well as in presentations, as the *Duquesne Revolution*.[7]

The major changes that were made can be broken down into two general areas:

- storage and reclamation of raw materials
- charging of those same materials into the furnace

Four furnaces of this size, producing nine hundred thousand total tons of iron per year, meant handling almost 1.5 million tons of 60 percent ore, 350,000 tons of limestone (at a rate of 0.25 tons per ton of ore charged), and about one million tons of coke based on 1,800 pounds (0.9 tons) per ton of hot metal produced. Considering that the Great Lakes are typically closed to shipping from November to April each year, room for about six hundred thousand tons of ore had to be provided in order to continue an uninterrupted operation of the furnaces through those months. This undertaking was no small task, anticipating that the plant would need to handle about 2.85 million tons of raw materials every year. Even that is quite an underestimate, since much of the ore had to be handled twice, once when it was taken from rail cars and put into the field for storage, and again when it was reclaimed for use in the furnace. Preferably, it would have been unloaded directly from the railroad cars into the supply bins at the furnace. However you look at it, this was an enormous undertaking.

The ore field was massive—1,085 feet in length, 226 feet in breadth, and 26 feet deep with a capacity for six hundred thousand tons of ore. Three huge electric-motor-driven ore cranes (usually referred to as an ore bridge) extended over the field, each with a clear span of 233 feet. There was an additional thirty-three-foot extension (for 266 feet total overall) on the furnace side in order to reach one of two rows of stock bins that were serviced by them. The clearance under the bridge's bottom cord was fifty-eight feet, slightly more than the maximum depth to which ore could be piled in the field. New all-steel construction, bottom-dump hopper cars, laden with ore from the lake port at Conneaut, Ohio (about 135 miles distant), would arrive on Carnegie's Pittsburgh, Bessemer, and Lake Erie Railroad.[8] Each hopper car would have been unloaded into stock bins, then transferred from a bin into a bucket on the storage yard side, which could then be picked up by the ore bridge, and finally emptied in the field. To reclaim ore, a scoop was dragged up the slope of the ore pile until it was filled with about five tons of ore. It was then loaded into a storage bin, unless the one where the bridge had stopped was full. In that case it was dumped into a railroad car to be taken to an empty bin. Each bridge was capable of handling between 1,500 and 2,000 tons of ore per ten-hour day.[9] The supply or stock bins consisted of two rows of bins, each running about the entire length of the ore yard, which was slightly longer than the layout of the furnaces themselves (see diagram in figure 7.1). The bins nearer the furnaces were primarily used to supply coke and limestone. The row nearer the field, which also contained two rows of bins, was for ore.

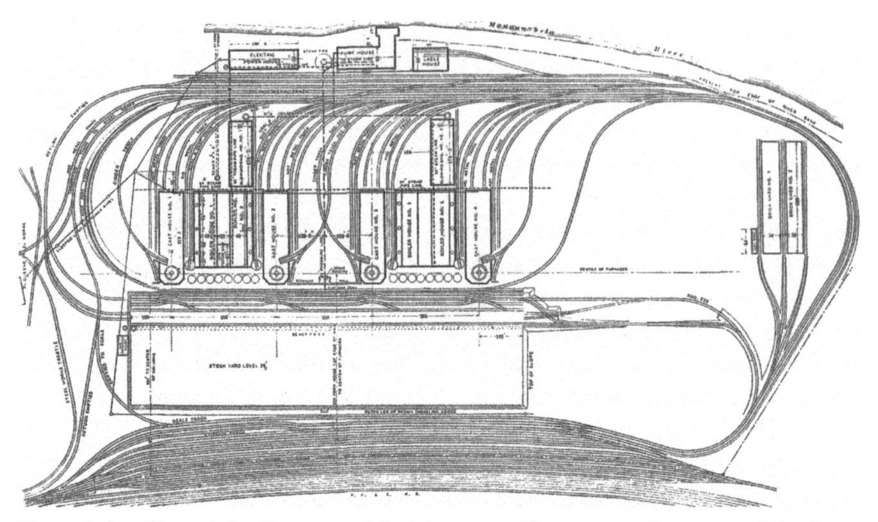

Figure 7.1. Plan of the Duquesne blast furnaces. The rectangular ore yard is above the maze of tracks at bottom. Atop that are the stock bins and above them are the four furnaces and their cast houses.

Railway and Engineering Review, April 9, 1897.

About 9,500 tons of ore, 3,600 tons of coke, and 2,200 tons of limestone could be stored in the bins. They each had railroad tracks mounted above to receive materials and tracks below to dispense materials, as required.[10]

To produce six hundred tons of iron per day in a furnace, it needed to be fed about 1,800 tons of raw materials of the quality available to the operators at Duquesne, when the plant was started. More than seven thousand tons were needed to sustain all four furnaces. In order to handle approximately three hundred tons per hour to feed the furnaces their daily ration, Marvin A. Neeland, the chief engineer for the works, designed a radically new system for hoisting material to the top in order to fill the furnaces. According to J. E. Johnson, a noted steel industry engineer, expert, and author on the subject of ironmaking: "When the Duquesne furnaces were designed . . . practically all the traditions of blast furnace design were discarded and the engineers struck out on lines boldly new in almost every direction."[11] The old vertical hydraulic elevator for lifting material was eliminated completely. The new system used a number of cylindrical buckets that were loaded with ore, coke, or stone from the rows of stock bins and transported on cars to the hoist house, where a long, steeply inclined bridge led from ground level to the furnace top. A steam engine and later electric-motor-driven drum with a cable that was strung up and over the top of the structure was then connected to a car on a track, suspended from the bottom side. One hoist was fitted to each furnace. When the hoist engine and cable drum were engaged, the cable attached to the car pulled it up to the top and later let it down the incline. The car had hooks on it that could grab a T-shaped bar projecting from the top of the loaded buckets, one bucket at a time, and hoist them to the top of the furnace (see figure 7.2). The buckets could hold either five tons of ore or two tons of coke or limestone. Cycle time to reach the top with the load, empty the bucket, and return to the bottom was a relatively swift one-and-three-quarters minutes.[12] Once at the top, though, the Neeland bucket was placed directly above the furnace bell and lowered by means of the contoured rails on the bridge on which the car traveled. When the hoist came to a stop, a portion of the carriage would continue to descend until the bucket also came to a stop against a gas seal, through the aid of a counterweight and cylinder.[13] But while the hooks of the car were still grasping the T-shaped bar, it would continue to descend, causing the conical-shaped bottom of the bucket, still attached to the T by the bar, to drop down, open the floor of the bucket, and evenly distribute the contents of the cylinder onto the large blast furnace bell immediately below.[14] The large bell could then be opened by remote control, allowing the load to be fed into the furnace at the desired time. There were no

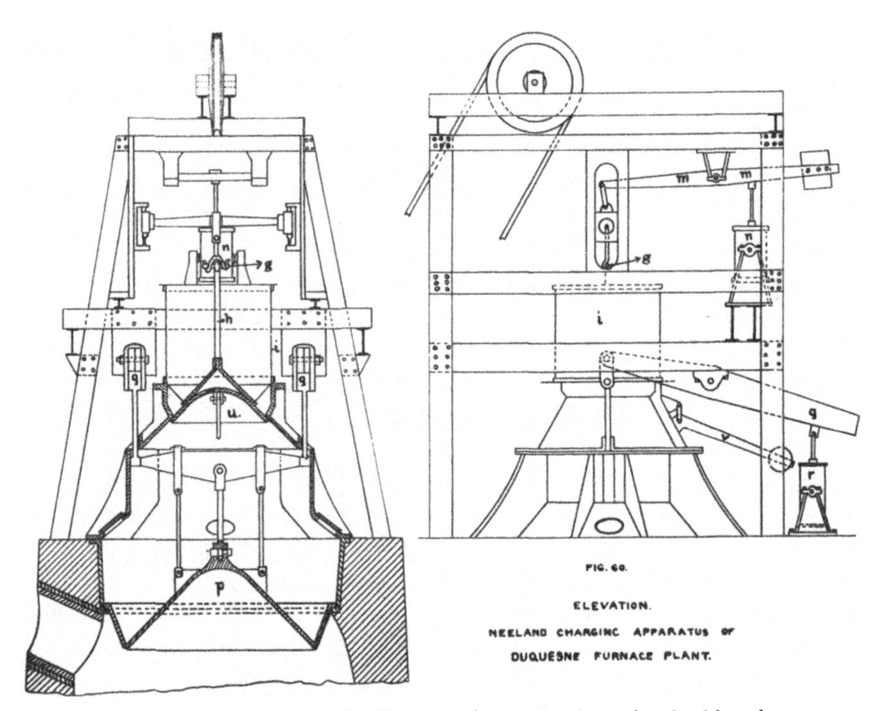

Figure 7.2. A cut-away schematic diagram of a Neeland top for the blast furnaces at the Duquesne Works.
An Outline of the Metallurgy of Iron and Steel, Hoffman, 1904.

Figure 7.3. View of the Duquesne Works blast furnace plant and ore yard c.1901.
Carnegie Library of Pittsburgh.

top fillers, because there was no longer any need for them. A single workman conducted the entire operation automatically from his location in the hoist house at ground level.[15] This system was so successful that frequently about seven hundred tons per day were produced, or about 15 percent more than the design predicted.[16] "This Duquesne furnace plant represents a very great advance on what has been done in America, and, of course there is no European plant that can compare with it in size and output. It has already gone right ahead of the Edgar Thomson furnaces [1897] and it is probable that it will improve on the present figures."[17]

Numerous people within the Carnegie Company, as well as at other companies, had been attempting to automate the blast furnace charging system for more than a decade prior to the inception of Duquesne. Most met with failure. In fact, what has been described as the first automatic skip-charging system was put into operation at Lucy Furnace in 1879.[18] Although this reference indicated that Lucy's skip began operation in 1879, a date of 1884 seems more likely.[19] Regardless, if Lucy's system wasn't the first, it was certainly a very early one. Skip charging, while similar in some respects to the Neeland system, was still significantly different. Skips are rectangular-shaped boxes (occasionally cylindrical or barrel shaped) that had fixed bottoms and were usually mounted on four wheels. The lading was both filled through and dumped out from the top of the skip. In order to empty, the skip had to be inverted somehow. This was usually accomplished very simply by having the top wheels of the skip follow the tracks at the top of the incline that then started downward (similar to a roller coaster at the top of a hill), causing the skip to pivot about the bottom wheels and invert and dump. Both with Neeland's plan and with skips, materials were run to the top of the furnace on an inclined plane, both pulled up there by a cable, and skips operated on tracks mounted to the top of the structure rather than being suspended from beneath. There were single skips and double skips. Double skips had three major advantages:

- They were twice as fast, since a loaded skip was going up as an empty one was coming down.
- As one was being dumped at the top, the other was being filled at the bottom.
- As both skips were connected by cable to the same drum drive, they balanced each other on the way up and down, which required less power.

Usually double skips rode on rails mounted side by side. These had been devised by William Rotthoff, sometimes listed as Rothoff or Rotoff, but the patent came under the control of and was modified later by Walter Kennedy, following Rotthoff's death.[20] There was also a design where skips were mounted on tracks above and below each other (E. G. Rust[21]). Rotthoff worked with Julian Kennedy at Lucy. He is named on the patent for the Duquesne charging system along with Neeland.[22] However, it might be true that skips had been used in Sweden some two centuries earlier.[23]

The only other successful mechanical charging system, besides Lucy's, that was still operating when the new furnaces were started at Duquesne was one known as a "Brown top" that was attached to a furnace located in Sharon, Pennsylvania.[24] The aforementioned Lucy skip had an "orange peel" type closure at the upper hopper on the furnace. It was recommended that it *not* be adopted at Duquesne.[25] In addition to these, there were a variety of other blast furnace tops developed in an attempt to be more compatible with the type of skip selected. To add to the confusion, there were double bell tops, single bell tops, and also annular bell tops (similar to one adopted and discarded at Lucy during the 1870s and at Edgar Thomson in the 1880s). Post Duquesne, there were McKee tops and Kennedy (Julian) tops, and many that were simply variations on a theme. The Brown top design that preceded Neeland had a single skip that fed a small, angled hopper that actually looked more like a chute, which was rotated part of a revolution after each skip load was dumped. This distributed the material more or less evenly in piles on the furnace bell before it was lowered to feed the stack.[26] Prior to Duquesne and even afterward for some time, with the exception of the Brown top and the one at Lucy, the myriads of combinations of systems and variations thereof simply *did not* work. Yes, raw materials were gotten into the furnace quickly, but it worked irregularly or developed numerous other operational problems with the furnace. Many of the managers in charge had little or no comprehension as to what was occurring with their automated systems:

> Many of them have tried out two or three, and others very many more types of tops . . . Many plants have adopted what is known as the plain double-bell style of skip filling, but many have had most disastrous results with it, and others after the most careful and conscientious trials to obtain an exact distribution, have reached the conclusion that no form or design of this top is satisfactory because the distance which the materials travel is effected by the size, shape and moisture contents of the different materials and by the velocity of the skip when dumping. What is perfect distribution on a dry day may be a frightfully bad one during a rain storm if the ore becomes wet, while a good distribution for coarse coke, free from fine material, may give horrible results with a fine coke much broken up.[27]

Moving away from hand charging was a difficult step for a number of different reasons beyond the large capital investment required. Men could see, think, and hopefully made good, honest, rational decisions that machines could not. Work that at first appears quite innocuous—for example, dump a load of material from a wheelbarrow into a hole—meant that top fillers had to be properly trained to do their job, because there was a logical progression that needed to be followed in order to keep the furnace functioning correctly. "Each barrow-load of material was dumped in a certain place on the bell, to be changed at the next charge. It was a common practice to charge coke, stone and ore in layers."[28] In other instances coke was loaded and dumped separately, while ore and flux were charged together. Much depended on local practices and preferences as well as the types of raw materials available. At some plants there was a cast iron box that contained burning coal or coke kept adjacent to the bell to ignite the escaping gas when the bell was lowered or to consume the gas that was almost always found leaking somewhere through a gap at the seal between the bell and the hopper. Burning it would help eliminate the possibility of men being overcome by the toxic nature of blast furnace gas.

> With the rapid increase in output which took place in the early nineties and culminated in the *Duquesne revolution*, [italics mine] the difficulties of filling the furnace became more and more serious. In addition to the vast difficulty of

> training and keeping crews of the size necessary to do this work by hand, and of supervising their performance of it to secure good furnace operation, it became almost a physical impossibility to handle the number of barrows required, and get them on and off of the hoist platform at the top and bottom of its travel, even though the operation was kept up at the highest possible pressure for twenty-four hours a day and practically without intermission. If any interruption occurred it became more and more difficult to "catch" the furnace after the stock line had gone down below the proper level; the fillers were kept almost up to their maximum speed to keep even with it under normal conditions and obviously had to make a stern chase in order to overcome any lead that it might get by an interruption in the charging.[29]

Human nature being what it is, even before the advent of the time when rapid hand charging became an absolute necessity, "too frequently, in order to save distance in wheeling, the larger part of the stock was dumped on the side nearest the hoist, or, to avoid a place where hot gas was escaping from a leak in the bell or hopper, preference was given to the cooler side." So, "irregular furnace work [was often simply] traced directly to dishonest top-filling!"[30] Justification for adopting mechanical charging was compelling for reasons far beyond the large increase in production or savings on labor cost, the most basic of these being the frequently fatal gassing and burning of men and/or projecting them into the yard from the top of the stack. "According to Mr. Frank Stewart 'about 60% of the fatal accidents of a furnace occur to top fillers, and the removal of the top-filler's position by the skip-hoist dismisses this kind of accident.'"[31]

A prime example of what could go wrong with improper filling is illustrated by the following:

> On August 20th [1895] a "hang" fell from the top of a furnace in the Edgar Thomson Steel Works in Pittsburgh, Pa . . . [Eight men] were killed, and eight other men received burns of such severity that some of them will probably die. It might be well to explain that a "hang" is a mass . . . adhering to the upper part of the blast furnace on the inside . . . [it sometimes] entirely chokes the upper part of the furnace and prevents the escape of the gases which are liberated in the reduction of ore. In the present case the "hang" had been neglected by the top fillers until the mouth of the furnace was nearly or quite choked up. A force of men was then sent up to remove it, and while they were at work it fell into the molten metal underneath, and the accumulated gases rushed out at the top, took fire, and exploded with a deafening roar. The sheet of flame which issued from the top of the furnace struck the men who were scattered about it, and blew them in all directions. One man was thrown over one of the elevators, and his body, striking a car below, was cut in two. Others were burned beyond recognition, and could be identified only by their clothing or by physical peculiarities.[32]

Eliminating the top fillers did not completely eradicate this type of accident, but it did diminish exposure to it.

A number of important conclusions can surely be gleaned from all of this. Primarily, it should be obvious that the proper distribution of raw materials in the furnace shaft was of fundamental importance to the successful uninterrupted operation of a facility regardless of the method(s) by which they were fed into the furnace. In addition, managing problems created by the segregation of these same materials became critical, if the move away from hand charging was anticipated. This was likewise true of developing an understanding of, as well as the means for dealing with, operating complications related to the fragility of coke. It also became clear that the double bell, or at least a system that provided a gas lock to prevent or to a large extent minimize the leakage of gas from the furnace stack when the big bell was lowered, was a necessity. The move to double bells saved gaseous fuel, estimated to be as much as 3 to 15 percent, which could be used elsewhere in the mill, Otherwise, it was wasted to the atmosphere when using a single bell.[33]

Problems introduced by the abandonment of hand charging and the adoption of mechanized material handling were numerous and multifaceted. To a limited extent they continue to plague ironmaking operations to this day. Volumes have been written on the subject. A number of these failings were directly related to problems associated with procedural errors made during hand charging, so the cause/effect relationship could be traced directly back to them.

In order to promote a better understanding for the causality of a variety of these failings, brief mention of a few of them seems warranted.

- *Segregation caused by the skip*: When dumping lump ore (and to a much lesser degree with Mesabi ore, which was a more fine material) from a single skip, large lumps having little resistance to flowing and possessing a higher inertia (high mass and speed relative to the fines), when being dumped, tended to end up on the side of the bell opposite of the skip. The finer material had more resistance to flow and moved slower and therefore ended up on the side of the bell nearer the skip. So by one simple step of the process, the ore charge (true as well for the coke and stone if their size distributions were similar) would readily segregate itself in this manner,[34] causing the furnace to have mostly coarse material on the side opposite the skip and a finer composition charge on side nearer. Because there were more gas passageways on the side of the furnace with the coarser materials, there was less resistance to gas flow through the furnace on that side. Since there were fewer obstructions than would normally be expected, if the entire charge were typically well mixed, even more gas than normal could and more heat did pass up through the shaft favoring that side. This would in turn cause the furnace lining and the shell to overheat high up on that side of the stack, because the cooling plates were limited to only the lower portion of the furnace. Overheating caused the brick to fail rather quickly. A sure sign that this had happened was the outside furnace shell plates visibly glowing red hot at night. Once this occurred, the furnace had to be taken out of service to make a repair or to reline.[35]

- *Segregation caused by use of a double skip*: Because the skips were side by side, they had to dump their loads into a wide hopper approximating the shape of a flattened funnel as opposed to one that was a smaller cylinder used with a single skip. This would compound the effects of segregation, first by emptying the skip as in the "single skip" scenario, and second from sliding down the slope of the hopper perpendicular to the skips to fill the central shaft above the small bell. Moreover, this effect was aggravated by alternately dumping skips on the left and then the right side.[36]
- *Effects of installing an improperly sized big bell*: If the size of the bell chosen was too large in relation to the furnace's top diameter, all of the material would pile up against the walls in the furnace shaft. If the bell was too small, it would pile up toward the center. The distribution in the shaft could also be modified simply by changing the location (the height) of the stock line in the furnace. If it were too high (too near the big bell), again the material will end up nearer the wall. If too low, it may fall closer toward the center. Many modifications were made in order to try to correct imperfections involving particular parts of a system, and although a complication of automation might appear insignificant on its own, the cumulative effect of these changes often led to compounding the errors in charging.

Another serious problem that arose from the adoption of mechanical charging was that dropping the coke during the process of transporting it from step to step in the material handling operation caused it to fracture. In old-style hand charging, the coke might be unloaded into a stockpile. When needed, the coke was forked from the pile by the bottom filler (using a tool that looked like a big pitchfork) and loaded into a coke barrow to be sent to the furnace top. Probably through an unintended consequence, using the fork instead of a shovel acted to screen out the smaller pieces of coke that fell between the tines. The larger pieces were forked into the barrow from little more than waist height at most. If it were loaded into the barrow from a storage bin, segregation of fines was similar to skip charging. Once it reached the top of the furnace, the top filler tipped the coke barrow forward and unlatched the front gate to dump it onto the bell. Again there was only a very short fall, followed by another drop into the furnace. It was a far different story with the double bell and skip system, however.

> Coke was dumped from hopper cars on high trestles into bins perhaps almost empty, involving a drop of 15 to 25 ft., dropped again from the gate of the bins to the bottom of the scale car or bucket, 6, 8 or 10 ft. more, dropped from the scale car into the skip, a distance equal to the last, from the skip to the receiving hopper some 7 or 8 ft., from the upper bell onto the main bell, about an equal distance again, finally from the main bell onto the stock line some 8 or 10 ft., if the furnace was full, up to any distance if it were empty.[37]

Many plants never took into consideration the general knowledge that coke was fragile and breaks apart easily, though it might have been easy to overlook that fact because the beneficial effects of hand selection and charging may have overshadowed that understanding. Coke is the scaffold on which ore and limestone rest, layer upon layer, as it descended the seventy-eighty-ninety feet to the hearth. It provided many of the passageways through which reducing gasses traveled upward on the way out of the furnace. If the coke was crushed too much, it would prevent the furnace from working due to impaired gas flow because of clogging by the fines throughout the shaft.[38] If it were segregated by the skip and bell, that could be a major factor contributing to channeling on one side of the furnace. So taken together, the interrelationships among the various members of a mechanical charging system can be quite complex indeed unless these problems were eventually overcome or the facility had to change to a different set of raw materials due to cost or availability. Many would later adopt some sort of intermediate screening at the blast furnace material-handling system to solve some of these issues.

The Brown top was successful to some degree, because it distributed entire skip loads of material into single piles on different sectors of the big bell each time that it dumped, in essence partially undoing the effects of segregation that occurred to some extent by forcing the skip charges to be deposited by the bell at an ever-changing location in the shaft.[39] Obviously, the degree of effectiveness of the Lucy skip must be questioned, since they recommended against using it at Duquesne. So, what if anything does understanding the idiosyncrasies related to mechanical charging have to do with the success of the Duquesne operation from its very beginning? For one, Neeland recognized the reasons for, and understood the consequences of, coke breakage that occurred during rough handling with automated equipment. He sought to correct the issue before the bucket system was ever put into operation. Even though there was still a large drop from the hopper cars into the storage bins, he minimized further breakage by using only a short chute or spout to load the bucket from the coke bins, and by locating the loading position of the bucket closer to the bin. Furthermore, since the bucket was taken directly to the top of the furnace from the bin without any further processing, using it also eliminated drops from the scale car to the skip, the skip to the hopper, and finally, the hopper chute to the small bell. Segregation of the material at each of these steps was eliminated, and the coke was evenly deposited onto the large bell from the bucket that was centrally located above it when it dumped.[40] Segregation was then limited to the chute between the storage bin and the bucket, totally eliminating all of the primary sources that had been recognized as failings in other systems.[41] Since breakage was not an issue with ore, he sought to diminish the effect of segregation of ore as well. To accomplish this reduction, he had the men load ore into buckets alternately from one side of a bin and then the other. Alternately destocking from a pair of bins directly opposite of one another effectively, if not totally, cancelled out many of the ill effects of segregation that would have occurred while loading from only one side of the bin. Again, by using the bucket all of the other typical segregation points were eliminated, because no transfer was made. Some years after the plant was in operation, a system of rotating the buckets on the scale car as they were being loaded was instituted to further eliminate problems.[42]

Experimentation and research into effective methods for filling blast furnaces was not solely limited to Duquesne, but was conducted elsewhere within the Carnegie Company and by other plants as well. According to Johnson:

> The first installation of skip filling commercially successful on a large scale was made at the Edgar Thomson Steel Works . . . By careful painstaking and long-continued experimentation with model tops built exactly to scale, and raw materials crushed and sieved to the same scale, the proportions were found which would give a reasonably uniform and satisfactory distribution of stock on the bell. A furnace was equipped with a filling apparatus of this design, and it was so successful that except for minor details nothing else has ever been used at that plant.[43]

Johnson's article was published in 1915, but in trying to establish an approximate time for when this equipment may have started operation, an article in a U.S. Steel employee publication from 1936 indicated that ET installed a double bell on a furnace in the latter half of the 1890s and used what ET claimed was the first automatic skip hoist in this country in August 1897. The article also stated that the first electrically driven skip was put in operation there on March 9, 1898.[44] The significance of these alterations was born out by the fact that in March 1905, four of the eleven furnaces at Edgar Thomson reached a world record production of 77,242 tons, surpassing the record output of the four Duquesne furnaces only the previous October.[45]

Early in the twentieth century Arthur McKee and Julian Kennedy would both design skip-fed blast furnace tops with automatic stock distributors. Kennedy left employment with the Carnegie companies in 1888,[46] and shortly thereafter he began his own engineering consulting firm in Pittsburgh. To a great extent he was responsible for the emergence of *hard driving* at Edgar Thomson. Later he was sent to take charge of the newly acquired and poorly performing Homestead Bessemer plant. He turned that facility around, transforming it into a world record holder when it passed ET. Later, he supervised the building of the first open hearth shop there as well. Kennedy was also dispatched to Lucy while at Homestead, where he revitalized that plant, essentially doubling production before he left the company.[47] McKee was born in State College, Pennsylvania, and was schooled there as an engineer. He worked for a few years at the Frick Coke Company and then both at the Edgar Thomson and Duquesne Works, during the Duquesne Revolution. When he left, he was next employed at Kennedy's engineering firm for about eighteen months, after which he moved to Cleveland, Ohio. Several years later in 1905, he began his own engineering company there.[48] Coincidentally, but not surprising, both charging devices were very similar,[49] but the McKee top was still being used in the United States at the end of the twentieth century.

Kennedy's quest for a new furnace-top design began in 1899 or 1900, not as a scheme to make and sell a new product, but to satisfy his innate desire to understand ways in which to prevent all of the havoc and mayhem that affected the men and machinery at the furnace top. Early bells and hoppers had been held in place by their own enormous weight. Later hoppers were bolted to a bracket that had been secured to the outer shell of the furnace. Explosion doors were added to the downcomers—for example, the main gas offtake pipes from the top of the furnace. Sometimes extra relief doors were added to the furnace top. During a severe slip, it was not uncommon for thirty to fifty tons or more of coke, ore, and limestone to be ejected from the furnace through these doors in less than a minute. Woe to anyone who was within the trajectory of these speeding meteors, or who had debris fall onto them in the yard far below. Kennedy rightly concluded that the outpouring of gas and ejecta that came from inside a furnace during a slip was in fact not as a result of an explosion, but more like the bursting of a balloon after the pressure was permitted to build to too high a level.[50] After performing some calculations, he was startled to find that there was not a single furnace in the United States "whose top would not be lifted by a pressure of [only] two and one-half pounds per square inch." Many could be lifted by much less.[51] He further concluded that the maximum pressure that could be expected during a slip was limited to the maximum pressure of the blast air that was produced by the blowing engines. Kennedy also determined that if he could build a furnace shell with the connection between the furnace and the top strong enough to resist about thirty-five pounds of pressure,[52] the size of the outlet piping could then be reduced, and less of the furnace contents would be expelled during a slip due to the increased resistance to flow by the smaller pipe. He dispensed with explosion doors whenever he could convince managers to do so, since they seldom properly sealed the furnace adequately and were often the *cause* of explosions by permitting outside air to infiltrate *into* the furnace.[53]

None of this is to suggest that actual explosions did not occur inside of the blast furnace. After all, the three prerequisites for an explosion—fuel, air, and a source of ignition—exist inside a working furnace 100 percent of the time that it is in operation. They were just not normally present in the necessary proportions. So furnace explosions did occur and do so to this day, even powerful enough to toss a modern furnace top hundreds of feet. Fortunately, these were and still are extremely rare.[54] In 1901 Kennedy introduced his new concept to the world on a furnace he had modified for the Iroquois Iron Company in Chicago. Reportedly, the furnace successfully met the test of heavy slips without exploding or ejecting tons of its contents. By 1907 there were twenty-four furnaces of that general design in operation.[55] The modern shape of the gas outlet piping in which the "gas is taken away from the furnace top by four gooseneck shaped downcomers, which unite downward into two self-cleaning gas mains conveying the gas into the dust catcher" seems to date from that time.[56] Based on the information presented above and in the preceding chapters, it should be no surprise to find that the evolution of modern blast furnace design has the imprimatur of the Carnegie Steel Company firmly and irrevocably stamped upon it, both directly through its accomplishments as well as indirectly from the experience and success of its alumni.

Although this was an amazing series of achievements, it isn't nearly the complete list of innovations instituted at Duquesne and its associated works. Electric overhead cranes of ten-tons capacity were installed in all of the furnace cast houses, and mechanical pig-breaking machines were used. The cranes would lift an entire twenty-six-foot-long sow, complete with pigs still attached (which looked like a huge comb), and move it to the pig breaker, where it was automatically reduced to more easily handled chunks that were loaded directly into cars at the breaker. Only a portion of the total daily production was cast into pigs (one estimate was about 620 tons c. 1897), as the majority of the molten iron was used at the plant's Bessemer facility.[57] Just like the top-filling operation, pig casting had evolved to the point where it was nigh on impossible for the men to keep up with daily furnace output at the larger plants when using traditional methods. At one-hundred-tons-per-day production rate of the later 1870s, there were roughly two thousand of approximately one-hundred-pound pigs to be carried by a crew of twenty every day, requiring one hundred trips per man per day from the pig bed to the railroad car. This was close to five tons per man, when the furnace was cast twice per day. It does not account for the additional work of breaking the pig or making up the bed. About ten years later, two hundred tons or more were produced per day, and casting from four to six times per day, a man was required to haul ten tons and more. It became necessary to begin breaking the iron and toting it while it was still hot in order to cast the furnace, break the pigs, carry the iron, and still have time to make up the beds. Imagine what it was like for a man when they were casting three hundred tons per day by 1890. According to Edward A. Uehling "The iron carrier must, in addition to the strength and endurance necessary simply to carry the iron, have a fire-proof constitution, to enable him to endure the heat, as well as the physical exertion."[58] It became more and more difficult to find men who were capable and more importantly willing to do this work. It was estimated that perhaps only 5 percent of all of the available men might have had these abilities. On occasion the men could not or would not be driven, refusing to do the work, especially during the hot summer months. Sometimes they would get drunk between casts. A blast furnace cannot be turned off like a light switch but continues to make iron until the hearth is full and there is no longer any place for the iron to go. It can be slowed, but the iron level in the hearth continues to rise until it reaches the tuyeres or causes a breakout of the furnace.[59] The tuyeres were made of bronze and would melt, releasing the water inside. The iron could freeze, or worse, there could be an explosion. Being unable to cast a furnace was a very dangerous situation, so serious that it was to be avoided at all costs. One simply cannot begin to imagine what the conditions would have been like if traditional pig beds alone were used when the behemoths at Duquesne began yielding six hundred tons and more per day, not from just one, but from all of its four furnaces.

Shortly after the new blast furnaces in Duquesne went on line, using new labor-saving devices added there to ease the demands of pig handling and breaking, they were quickly superseded by a new invention that had just been developed at Carnegie's Lucy Furnace. James Scott, the superintendent at Lucy, had been conducting experiments for some time in order to eliminate pig beds altogether. Initially, he cast the furnaces into ladles that were then taken to a water-filled pit into which a steady stream of iron was slowly poured, and where a submerged conveyor was used to continuously remove the chilled nodules of iron from the water and load them into a waiting car. The quality of the iron produced was poor, and the maintenance of the machinery proved costly. So he next succeeded in casting pigs into moulds submerged under water, progressing to filling moving moulds slowly as they passed under a direct stream of iron from a ladle. After advancing to this point in his experiment, Scott soon discovered that there was a preexisting patent from late 1895 concerning this process that had already been granted to Uehling. A major stumbling block in machine casting was a problem with newly cast pigs sticking in the mould. This would then cause the iron to splash and spill all over the machinery the next time that particular mould came up to be filled again, sometimes destroying the machine. Uehling solved this by spraying each mould with a refractory, in this case a lime slurry, which acted to release the solidified iron pig from the metal mould at the end of the conveyor. Reapplying the lime spray was necessary each time a mould passed by on the bottom of the continuous loop of two heavy chains that suspended the moulds between them. When the mould was returned to the top, it was filled with molten iron again. The Carnegie Company purchased an option to build one by Uehling's design for a test, and it promptly secured the rights to develop the machine. It took two iterations to sufficiently work out the bugs and get the machine to run satisfactorily at Lucy in March 1897. From then on, the total 750 tons of output of the plant was cast by machine.[60] Uehling credits both Scott and the Carnegie Company for the practical development of this concept.[61] Soon after its demonstration at Lucy, a second machine was built for the Carrie Furnace Company. In short order the Carnegie Company bought the two-furnace plant (The Pittsburgh Furnace Company) from the Fownes Brothers to help supply the Homestead Works, which had no blast furnaces of its own at this point. That same year, 1897, pig casters were installed at both the Edgar Thomson and the Duquesne Works.[62]

As previously mentioned, since not all of the ironmaking capacity could be used at the Duquesne Bessemer plant, a procedure was developed by which the remainder of the hot metal was transferred to the Homestead Works, whenever practical, to feed the open hearth furnaces there, saving the energy required to remelt cold pig from the machine. Coke dust was layered over the hot metal in open-topped ladles to insulate it during the approximately four-mile trip by train down the Monongahela River. The coke dust covering did no harm to the metal quality (see figure 7.5). The coke dust was otherwise a waste product. Machine cast pig offered a considerable cost savings (at least $0.25 per ton) over sand, when used in a basic open hearth, due to reduced flux requirements, less slag generation, and less damage to the basic furnace's hearth.[63] But using the hot metal directly saved even more money in fuel costs required to melt pigs.

The entire list of innovations made at Duquesne has still not yet been covered. The plant had an extensive facility for generating the electric power needed for the motors on the cranes and ore bridges. In addition to the power for the cranes, the works had a total of six sixty-light, direct-current generators for arc lighting, as well as a forty-five-kilowatt alternating-current generator to light nine hundred incandescent bulbs.[64] A sixty-light generator refers to one that is capable of powering sixty individual arc lamps, each one suitable for producing a sixteen candlepower light output.[65] "One of the most complete electric power equipments in the world is that which has been put down at the Duquesne Steelworks and Blast Furnaces, Cochran, near Pittsburgh."[66] The massive undertaking of improvements just described was of extraordinary proportions, and they were inaugurated at the Duquesne plant by the Carnegie Company, despite the American economy being enveloped in a severe business depression during 1893 to 1896.[67]

Dispensing with top fillers, bottom fillers, pig bed laborers, and using the new all-steel, bottom-dump railroad hoppers eliminated hundreds of men from the workforce, while the few people needed to man the hoist houses, ore bridges, pig caster, and scale cars for bucket loading required the addition of only several dozen or so of new employees.

Eventually, after the complications of the double skip and double bell charging systems had been worked out sufficiently (see figure 7.4), the Neeland system fell to the wayside, because it had become the more intricate of the two material handling processes. However, by this time Carnegie was gone from the business.

Figure 7.4. View of the Edgar Thomson Works blast furnace circa 1905. The skip inclines for the loading system developed by ET are located on the opposite side of the furnace. Their system used a double skip and a double bell and eventually superseded the Neeland system of bucket charging. Lock and Dam #2 on the Monongahela River are foreground of the furnaces.

Library of Congress collection.

Figure 7.5. A train load of open top ladles full of molten iron covered with an insulating layer of coke dust, are likely bound for the open hearth furnaces at the Homestead Works. The Carrie Furnaces are across the Monongahela River and are the probable source of the hot metal, c.1905, supplanting the earlier sources of hot metal from Duquesne or Edgar Thomson.
Library of Congress collection.

During 1899 the demand for open hearth steel was ever increasing, so in the fall of that year an investment in a new shop to take advantage of the bullish market conditions was built at Duquesne, going into operation in October 1900. It consisted of twelve fifty-ton furnaces (by 1903, fourteen), all in a single line. Fifty tons was considered the largest practical size for stationary furnaces at that time,[68] although a seventy-five-ton moving furnace had already been built at Pencoyd for a Benjamin Talbot experiment, and a two-hundred-ton furnace of that type was being contemplated at J&L, but these were for a special process. The new facility was provided all the latest modern equipment for making steel:

- seventy-five-ton capacity electric overhead-traveling cranes in the pit for teeming the finished steel into ingots
- pouring platforms for casting ingots on cars in the open hearth shop
- means for hydraulically stripping the moulds from the ingots in a separate adjacent building

- additional forty-ton overhead cranes were on the charging floor for handling the ladles of molten iron, among other things, that were delivered at one end of the shop for loading liquid metal into the furnaces
- three Wellman design charging machines to load the scrap, ore, and limestone
- a number of small five-ton electric traveling cranes for handling scrap and other materials for filling the charging boxes, which were weighed afterward

The material handling was conducted in an area adjacent to the steelmaking shop. The main building that housed all of the furnaces was well over nine hundred feet long. There was an additional structure that housed a "skull cracker," a large crane mounted fifty feet above grade that could drop a heavy weight a long distance (therefore at high speed) onto large pieces of scrap to break them into more easily handled sizes, somewhat like a child could do to a tea cup with a hammer. This essentially eliminated the need for having furnaces with movable roofs like they had at Homestead to recover large pieces of scrap. The furnaces and charging floor of the new shop were elevated nine feet above grade, a departure from the ground-level layout of the three preceding shops at Homestead. It took advantage of all modern electrically driven equipment for hoisting materials, an idea that should have been, but was not realized, when the #3 open hearth shop was built at Homestead only two years before. Remember, the Homestead furnaces were built at grade. Other steel companies had used elevated charging floors and furnaces for some time. It was almost a necessity at plants with tilting furnace designs. Circular pits were still used on the pit floor at Duquesne, where the ladles were filled with the molten steel, but these were only slightly more than three feet deep, instead of the nine or ten feet at Homestead. The shallow pits did serve to collect slag that overflowed the ladle at tapping and made it easier to clean. For a while the company tried using metal hooks placed in the ladle pits that were later removed by a crane for hoisting large pieces of slag once it cooled enough to freeze to the hooks. Kloman developed the idea during some happier times at Lucy Furnace, but that might not have been the plan for Duquesne. Natural gas was the fuel of choice for the Duquesne furnaces, but a contingency plan for gas producers was made in the event that a shortage developed for the preferred fuel.[69]

In addition to the material handling improvements made at the Duquesne furnaces, G. H. Hulett of Akron, Ohio, devised a wholly new method for unloading ore from boats arriving from the company's Lake Superior mines.[70] Following a brief experimental trial of the original unloader concept at the company dock in 1899, the new year found the installation of three machines of improved design at the Pittsburgh and Conneaut Dock material handling facility, located at its lake port in Conneaut, Ohio. The Carnegie Company had a total of four operating there by 1901.[71] Arthur Johnson stated that this invention was so important that:

> The greatest revolution in ore handling that has yet been made will, without doubt, be brought about by the Hulett ore unloader . . . With all other ore unloaders the buckets are filled by hand, and as the work is very heavy, only the strongest men have endurance sufficient to enable them to work in the close holds of the vessels.[72]

The existing conventional equipment used buckets of the hand-loaded design, which held about one ton or slightly more, depending on the type of ore being handled, soft or hard.[73]

The new Hulett unloader used a "clam shell" style bucket attached to the bottom of a stiff leg that was constantly held in the vertical position at the end of a "walking beam" by means of a parallelogram configuration of the beams in the structure. The bucket had a capacity of ten tons per bite.[74] A Hulett was designed to move in three planes—vertically, horizontally (the width of a boat), and the entire length of a ship. In addition, the bucket could also rotate. They looked like monstrous dinosaurs that weighed about four hundred tons and towered fifty-five feet into the air above the dock.[75] The heavy weight of the equipment was intentionally kept unbalanced on the leg portion in order to provide a downward force to secure more easily the loading of the clamshell bucket as it closed around the ore. No men were needed to load the bucket, with the exception of when they were required to clean the hold of the ship of the 5 to 10 percent of the ore that could not be reached by this style of unloader. Each new machine could empty the ore boats at a rate of 250 tons or more per hour. And when used in concert on a single vessel typically holding four thousand to eight thousand tons of ore,[76] the four machines could unload it at a rate estimated to be between one thousand and twelve hundred tons per hour. Their installation was believed to have tripled the capacity of the dock.[77] The original Huletts were steam powered and required a crew of three or four to operate: minimally an engineer, a fireman, and the operator, whose cab was located in the leg of the walking beam above the bucket. This last man actually entered the hold of the ship inside his control pulpit, and he could see the ore just slightly below him, as the bucket closed on it. About four men were used in the hold to clean and load the bucket with the remainder of the ore that could not be conveniently reached. Soon after, electrically driven Huletts required only a single operator (see figures 7.6 and 7.7).

Although not obvious, the choice of location for the dock at Conneaut was a scientific decision made by the Carnegie Company. The site chosen was based on finding a good harbor near the longest water haul for the ore from

Figure 7.6. The Carnegie Company's three steam driven Hulett unloaders nearing completion of construction at the company's Conneaut, OH dock facility in 1900. The fact that they are steam driven is apparent from the piping on the machine and the large cylinder mounted on the top of the walking beam.
Detroit Publishing collection, Library of Congress.

the port near the mines that was simultaneously the shortest rail haul from the receiving dock to the mills in Pittsburgh. This was due to the fact that water is nearly always the lowest-cost route for high-volume, high-tonnage freight traffic.[78] It was estimated that one Hulett machine could replace at least six of the older kind, and that the cost of unloading ore with them, including the cleaning of the hold of the boat, was about three cents per ton, compared to the cost of shoveling, valued at fourteen cents per ton with an additional 0.7 to 1.37 cents for handling the buckets when using other types of machinery.[79] So there was a significant cost advantage when using a Hulett unloader as well as a large reduction in unloading times for each boat, and consequently the improvement in ore dock capacity, formerly described as being in the range of 300 percent. Previous methods required a large force of men and

Figure 7.7. A view of a more modern electrically operated Hulett unloader c.1942 at Cleveland, OH shown here to illustrate the leg and bucket structure of the machine. The operator's cab (square opening) is located immediately above the bucket in the leg. Library of Congress collection.

took between ten and twenty hours to unload a boat.[80] These improvements perfectly complemented the recent developments made to the ironmaking facilities at Duquesne. In addition, this system was augmented by the installation of elevated dock(s) equipped with a series of pockets and chutes that facilitated the rapid discharge by gravity of seven thousand to eight thousand tons of newly mined ore into the boats. Loading at such docks required approximately three or four hours at the Lake Superior end.[81]

Hulett unloaders lasted for more than a century; the final two were operated for unloading coal until the end of 2001 by the LTV Steel Company in South Chicago. Their ultimate demise resulted from the development of self-unloading ore boats in the later part of the twentieth century and the subsequent reduction in cost of unlading together with the bankruptcy and permanent closure of the plant owning the last two existing units. None remain. All have been disassembled.

Part IV

ANALYSIS

Chapter Eight

Summation

> In ten years after we began the manufacture of steel rails in commercial quantities, which was in 1867, the charge for transporting a bushel of wheat by railroad from Chicago to New York was reduced from 44.2 cents per bushel to 20.8 cents, and it has since been further reduced to 8.75 cents [1904 date of this article]. In 1860, with only iron rails, the charge for moving a ton of freight one mile on the New York Central Railroad was 2.065 cents; in 1870, after we had commenced to use steel rails, the charge was reduced to 1.884 cents; in 1880, when steel rails were in more general use on our trunk railroads, the charge was further reduced to *8.79 mills* [italics by the author], and in 1901 it was still further reduced to 7.4 mills.[1]

One thing is certain, although it is contrary to what many people choose to believe, the vast fortune amassed by Carnegie was not won through reducing the wages of his company's workers. A simple calculation will bear this out. In 1900 for instance, the last full year for which profits were recorded for the steel company, they amounted to $40 million. Although this is the figure routinely reported, Charles Morris in *The Tycoons* argues convincingly that the figure was nearer to $30 million.[2] Employment at the three steel-producing plants in Pittsburgh, including those making iron at the Lucy Furnace, was about fifteen thousand men. At a reasonable estimate of $3 per day average earnings per man or about $20 per week, this amounts to about $16 million total annual wages or about 40 percent of all of the profits reported for that year. At the $30 million figure, this was slightly more than half. Most of the men would have had to work nearly for free before attaining a significantly larger impact on those profits. This did not just happen accidentally. This did not occur because Carnegie cut wages by a few dollars per week. Obviously, for the vast majority of years when the reported profits were a mere two or three or more millions of dollars, this wage-and-profit relationship did not hold, but neither was the total employment as high.

Carnegie's large fortune was based on the appraisal and sale of an asset whose fair value exceeded by many times all of the profits for all of the years that the steel company was in existence. The reason for this success was primarily technological in nature and was directly dependent upon the innovative equipment and practices developed at Lucy, Edgar Thomson, Homestead, and Duquesne. It was also due to an enormous rate of capital reinvestment in the company. It is abundantly clear that the evolution of the American (and to a lesser extent the world's) steel industry closely parallels that of the Carnegie Steel Company, not vice versa. Building upon the basic principles of iron and steel making devised by others (mainly in Europe), through continuous innovation, automation, and reinvestment the Carnegie Steel Company reshaped the entire industry in such a way that in order to remain competitive with his company, his rivals eventually had to conform to the standards and ideals of Carnegie or face failure.

With our contemporary picosecond-speed computers, microchips, the Internet, interplanetary spacecraft, and the like, it is sometimes difficult to wrap one's mind around an era like Carnegie's when *gorilla men are what we need* seemed to be the standard. For that reason it might be somewhat difficult to comprehend that in numerous instances his was the cutting edge of technology for the industry and in some cases far beyond the cutting edge. Carnegie and his company applied science and powerful reasoning to gain an in-depth understanding of the production process at a time when many others formed the data to fit what they believed, rather than conforming their beliefs to fit what was revealed by the data.

The Carnegie Company's production statistics became so phenomenal that they were sometimes reported in books and journals in England and elsewhere as a percentage of the total output of other *nations*. These figures were not small. For instance, in 1898 the company's production was reported as 56 percent of the total output of the United Kingdom. When considering the production of open hearth steel versus that of Germany, Belgium, and France, it was greater than the combined product of all three![3] In 1901, *Iron Age* reported that for October of that year the rate of production of the Carnegie Company exceeded 40 percent of the output of the United States for ingot steel (both Bessemer and open hearth) when compared to the same period in 1900. In the same article a list of twenty-four new monthly production records was presented for its various plants and for particular mills within those facilities.[4] Other data from 1898 provides a glimpse of this prowess, when it was stated that this single company with only three integrated plants in Pittsburgh made 5.34 percent of the world's iron and 9 percent of the world's steel.[5]

IRONMAKING

For each of the three consecutive decades that it was in business, the Carnegie Company reinvented the iron smelting industry to the amazement of the ironmaking gentry, often producing what knowledgeable men believed was unobtainable. From the inception of operations at Lucy in 1872, it took only two short years to become the world's iron production leader and the first anywhere to produce one hundred tons in a single day. Through the unconventional use of a chemist, the importance of analyzing raw materials became readily apparent. This went against the grain of traditional "seat of the pants" or "rule of thumb" producers. Yet when the subsequent application of those and other scientific principles were embraced, it proved to be a savior of the plant. As Carnegie described the inconsistency in ore supplies that were available to Lucy, enlisting the expertise of a chemist demonstrated: "The good was bad and the bad was good, and everything was topsy-turvy. Nine tenths of all the uncertainties of pig-iron making were dispelled under the burning sun of chemical knowledge." Benefitting by this knowledge, high-quality ore was secured at low prices, because the general opinion was that those materials available to Carnegie at the time were generally thought to be of a lower grade. Adopting a simple procedure like counting and maintaining the rate of revolutions on the blowing engine proved a revelation and introduced consistency to the furnace operations. Although not originating there, and acquired at no cost, the importance of the relationship in maintaining the *steady flow of air* to the *steady flow of raw material* was recognized, so the practice was adopted, and the pressure-based control of the blow engines was abandoned.[6] Lucy was also an early, if not the earliest, proponent of skip loading furnaces, bringing Lucy the recognition on the world stage it deserved.

As the new blast furnace department for Carnegie emerged at the Edgar Thomson plant in the early 1880s, the use of air was again reanalyzed, and overblowing a furnace ("A") with more than twice its cubical capacity every minute began. No one had ever done this before. Furnace output under the new parameters was phenomenal, and by the end of the decade, output on the best furnaces was more than triple that of Lucy's 1878 iron production, while using only a minimally larger furnace (measured by internal volume). Of course, this first required building a facility that was capable of consistently blowing and heating such large volumes of air and was not merely an accident. Obviously, there was at least some forethought. Steel industry veterans even contrived a new term to describe this phenomenon, calling it *hard driving* or *rapid driving*. The new term was sometimes twisted when used by the nonbelievers in an attempt to ridicule the coincidently heavy wear on the furnace lining brought about by these practices. Air flow was not the only property that was a focus of experimentation. Air temperature was varied both higher and lower to observe the effects, and furnace stack dimensions (ultimately those pertaining to the internal shape and not the height) were adjusted to analyze their ramifications. When changes were made to the lines of the furnace stack, the plant had to suffer the effects of poor decisions for months. Yet within a decade the daily hot metal production on the best furnaces rose from 132 tons to 310 tons, while simultaneously, the fuel rate dropped 33 percent from 2,850 pounds to only 1,920 pounds of coke per ton of metal produced. In other words, more was made from less. This was accomplished while lowering the earlier blowing rate of about two times cubical capacity downward to around one-and-one-quarter times (see table 5.9 and also chart 5.1 on improvements made to iron production at selected Pittsburgh furnaces from the same table); as a result, reduced, among other things, was wear and tear on the blow engines, the physical facility required to generate the amount of steam demanded to drive them, as well as the fuel needed to make the steam to power them. By the end of the 1880s era, the United States moved ahead of the British to become the leading iron producer in the world. A large part of that success is due to the effort and genius of the men at Edgar Thomson.

At Duquesne, the problems involved with automatic feeding of the blast furnace stack in relation to the problems of proper raw material distribution were solved—nowhere near perfectly, but in sufficient detail to become easily the most productive plant in the world. It was not a hit-or-miss success, but it was planned from the inception of the idea to the realization of the facility to produce 50 percent more iron from a furnace than anyone else in the world had ever done. The plan not only met but also exceeded expectations. The intended result could not have been accomplished through a simple "build a bigger furnace" approach. It is important to note here that the people with the greatest iron outputs against whom the Carnegie Company was primarily competing were in fact *themselves*. The Neeland system that satisfactorily solved the distribution problems was a clear, yet subtle, departure from skip loading. The enormous material handling facility (i.e., the ore yard, bridge cranes, and storage bins, etc.) required to support new production goals was a bold commitment on the part of the Carnegie Company. Remember, these expansive innovations were embraced during a financial depression. The world of steel demonstrated the depth of respect and level of esteem in which these men of the Carnegie Company were held for what had been accomplished, when this new advancement was named the Duquesne Revolution. Further progress was forthcoming from Lucy: overcoming problems involving the elimination of pig beds, and developing a practical form of the Uehling pig-casting machine, which was adopted throughout the company.

Edgar Thomson, notwithstanding the progress made at the Duquesne plant, continued to develop further understanding of raw material distribution in the furnace stack through experimentation and analysis of operating scale models of the plant under differing charging scenarios and conditions. Soon a world record was surpassed that exceeded Duquesne and the Neeland system with their simpler double skip and double bell configuration. This was subsequent to the sale of the company to the Morgan interests.

Carnegie-owned limestone quarries, Lake Superior iron ore mines, and coking coal properties provided materials that were among the best quality available. Some added benefits were derived from beyond the scope of the improvements carried out in the Pittsburgh district. These materials moving from mines to the docks were a significant complement to operations in the Mon Valley. Items such as:

- the invention, installation, and expeditious development of the Hulett unloaders
- the consequential realization of the speedy recovery of ore from ships at the Conneaut docks
- the more-or-less simultaneous installation of rapid-discharge ore docks for filling boats at the Lake Superior end of the supply chain that significantly reduced the costs of handling while increasing the capacity of these facilities at the same time
- the adoption and use of rapid discharge (for the time), bottom dump, all-steel construction hopper cars for transferring ore from the lake port to the mills in the Steel City. Earlier methods of emptying cars depended on hand shoveling the ore through a trap door in the floor or even over the low walls of the car.

All of these new ideas hastened the transfer of ore, reduced the costs of doing so, and increased facility utilization, which eliminated the need for capital investments for more or larger operations to do the same at these locations. In addition, the Carnegie Company purchased and then revamped the nearly nonexistent right of way of the newly named (Pittsburg) Bessemer and Lake Erie Railroad (B&LE), which operated between the Conneaut dock and Bessemer, Pennsylvania, several miles north of the mills in the Mon Valley. It also built the Union Railroad, a switching line, to interconnect the Homestead, Duquesne, and Braddock works and connect them with the B&LE at its southern terminal. This way it could not be held hostage by other railroads for delivery of iron ore, one of their most important raw material supplies. In addition, the steel company started the Pittsburgh Steamship Company to move ore from its mines in the Lake Superior region to the dock at Conneaut, maintaining complete control of this critical commodity from the time it was removed from the earth at the mine until it arrived at the mills in Pittsburgh. Today we would call this logistics.

BESSEMER STEEL

The Edgar Thomson Works in Braddock was the first mass-produced steel facility in Pittsburgh. The region had relinquished its superior role as the nation's premier wrought iron and crucible steel maker to the newer Bessemer works of the Troy, Pennsylvania, Cambria, and other steel companies. When it went on line in 1875, the ET Works was merely the eleventh among similar plants built in the United States. If you use that as a criterion to measure importance, ET was not very significant. Two of those eleven plants had already been abandoned, and according to the *Directory* for 1876, most of the others had two five-ton vessels, the same as installed at Braddock. There were two exceptions: the Cleveland (Newburg, Ohio) Rolling Mill Company, which had four five-ton converters, and the old Troy works, which had one additional furnace of two-and-one-half tons capacity.[7] All in all, excluding the Cleveland plant, everyone was pretty even when comparing the projected capacities of facilities.

As indicated in chapter 5 on Bessemer steel, per plant production in America was steadily increasing. Holley had devised a new, very efficient facility layout in addition to raising the general furnace elevation and inventing a practical form of changeable furnace bottom. He also switched from reverberatory furnaces to using cupolas for melting the pig for the charging metal after first making several modifications to increase their yields. When the Edgar Thomson plant came on line, it swiftly became one of the foremost producers of steel, similar to what had occurred at Lucy in the ironmaking arena, soon averaging almost 1,300 tons per week over a fifty-two-week year. Numerous indicators show that the ET Works was often the country's leading producer, at least as far as two vessel plants were concerned. By 1878 it was considered the leader when compared to similar-sized plants anywhere in the world.[8]

The Holley improvements predated the construction of the Braddock works, and so were implemented there as well. Most, if not all of the earlier shops designed by him, were also similarly equipped. Since Carnegie's furnace output quickly exceeded the capacity of the original rail mill to roll all that could be produced by the plant's Bessemer converters (witnessed by the previously mentioned sale of billets on the merchant market prior to 1880), something other than the Holley modifications must have been responsible for Edgar Thomson's success. They were routinely making three thousand tons in a week by 1880, and by 1890 they were well over seven thousand tons with a three-vessel plant. Although they were in the forefront, some of the other American producers like Union (Chicago), Cambria, and South Chicago were close on their heels, sometimes ahead. The Cleveland Rolling Mills, which earlier was the factory with the most furnaces, opted to begin open hearth steelmaking with two furnaces in 1875, and in the process two of the four converters were dropped. Though it had five open hearths by 1886, it soon began to fade into the background.[9]

Edgar Thomson produced rail almost exclusively until about 1890. A large part of its success was due to the management of Jones. Ramping up production was in large part due to ramping down the cycle time on the Bessemer converters. As Howe intimated, fast production was more connected to what you did "between the blows" and not as much during. Probably the most important thing that Jones did was to convince Carnegie to permit him to move away from working two twelve-hour shifts and institute a three-turn, eight-hour workday in 1878 without any diminishment in wages to the labor force. Men could work harder and more consistently for eight hours than twelve. They had better opportunity to rest, and therefore they produced more in less time. Jones attained a certain stature in the eyes of his workers, so they were more likely to do things that normally they would have resisted. The "broom award" was also a striking bit of genius on his part. It instilled a competitive spirit in the men, made work more playful, and established a feeling of camaraderie, all for no cost. In 1888 when Carnegie forced Jones to put the men back on twelve-hour shifts with no increase in wages for the extra hours of work, I believe that Jones retained his status with the men. This idea is supported by the fact that shortly afterward, upon Jones's unexpected death late in 1889, the men had considered a work stoppage until after he was buried before they agreed to stay on the job. An estimated five thousand people or slightly less than double the entire workforce at the plant attended his funeral.

No other major steelmaker adopted the eight-hour day. This was another instance where Carnegie's desire to reduce costs overrode his self-proclaimed commitment to being the friend of the workingman. He somehow used the reality that since others had not followed his lead, it was permissible for him to return to the status quo. Had Frick been the overriding factor in the reversal, little would have been said, because it was well known that he "[regarded] men as a commodity like anything else used in manufacturing—something to be bartered for as cheaply as possible, to be used to its utmost capacity, and to be replaced by as inexpensive a substitute as was available."[10] Carnegie got into trouble because he tried to work both sides of this issue.

Jones also improved, invented, and adapted devices and/or procedures that increased throughput, reduced fuel consumption, and/or minimized waste. For instance, he patented a revised Holley vessel bottom and added a more convenient linking system, making it quicker and easier to change converter bottoms. Holley himself lauded these revisions in print.[11] Jones patented a method of adding steam to the blowing air to reduce high bath temperatures and to eliminate the time required to add the scrap needed as a coolant.[12] He added his own design of hydraulic lifting tables to the rail mill to address a bottleneck, eliminating two crews of men and improving safety, since the workers were in physical danger and could not keep up with the rolling schedules in large part due to poor working conditions. ET adopted the use of hot metal directly from the blast furnace for the converters (a European development) only months after it was first applied in America at South Chicago. This saved the need to cast pigs and then remelt the iron in a cupola. Only facilities that had the best (most stable) blast furnace operations could use this method without suffering extensive production problems at the converters.

In 1889 Jones invented and patented the "mixer," which blended out the inconsistencies found in the iron's chemistry (mostly silicon and sulfur) from cast to cast and dispensed with the routine need to cast into pigs or to use the molten metal directly as cast from the blast furnace. Although it was superior in many ways to pig casting and cupolas, use of the direct metal was never totally free from creating problems at the converters. The mixer essentially eliminated them. The difference between using the direct method and the mixer might seem almost insignificant, because an increase of silicon (or decrease) in the iron could be easily compensated for by simply blowing the heat slightly longer (or shorter for less). The problem was the unpredictability of that variation. It was understood that it was eliminated within a matter of seconds of being there, and as has been previously stated, if blown only seconds too long, the steel quickly became overoxidized, and if blown only seconds too short, the steel could be hard and brittle. Consistency was the key.

That same year Jones also patented a horizontal ingot stripper. The original intent was to remove sticking ingots more easily from the mould, but it was soon discovered to speed up the blowing cycle since it was learned that the ingots and the ingot moulds could be removed from the casting pit as a unit. Both were then placed on a car as a unit and transported to the stripper for separation. This eliminated the time involved with the two-step process of stripping the mould and then removing the ingot before placing a new empty mould in the pit. There was, however, only a short-term gain after it was implemented, because it was immediately superseded in 1890 by a new procedure of casting steel into ingot moulds that were already mounted on rail cars. First tried at the Maryland Steel Company (although previously suggested elsewhere), it was quickly adopted at the Edgar Thomson Works and is still used in some plants today.

Jones devised and patented another method to increase yields. It was something he called "compressed steel." By using a specially designed mould, stool, and cap, he applied steam under pressure (at 80 to 150 psi depending on the type of steel) to the top of the liquid steel in the mould, causing a crust to form on top of the metal and preventing a cavity called a "pipe" from forming in the ingot. The end of the ingot with the pipe normally had to be cut off and scrapped; yields were said to increase by 2.6 percent after application.[13]

Jones used the accounting system developed by Shinn as a tool, first to isolate and then to pinpoint sources of trouble within the process. He could then use his own mechanical prowess to address weaknesses in the equipment and apply his managerial skills if the problems were in the realm of human relations. Shinn's accounting system was an outgrowth of one he had devised from his railroad days. At first it was unique to the steel industry, giving the Edgar Thomson Steel Company insight into what they could and could not do as far as pricing was concerned, and provided more accurate answers to any situation. This is not a complete list of improvements made by Jones, but it should give you the picture.

With Carnegie's absorption of the Homestead Works in 1883 and the Duquesne Works in 1890, he then had three separate Bessemer shops in Pittsburgh. The two new facilities were afflicted with both labor and management problems at the time of purchase. Neither plant could make production goals or profit. Part of their problem was Carnegie's working against them in the market and in the rail pool. Most important, though, was labor's constant threatening to and actually going on strike, as well as management's constant ranting against unions, inciting them to strike. Once Carnegie management was in place at Homestead, in little more than two years, Julian Kennedy brought them to producing more Bessemer steel than they had ever before, and a year after that they surpassed ET, gaining a world record. Once the plant came under Carnegie control, making procedural changes to cast larger ingots helped them. This was possible due to the fact that a stronger rolling mill already existed there. So the plant could produce different product lines such as structural shapes and plate instead of the rails, which at that time for various reasons still required smaller ingots. Rail production was shifted to the more efficient Edgar Thomson operation.

The Duquesne Works accomplished similar feats when Carnegie's managers took control. Without going into the details of it, in October 1901, slightly more than ten years after they were purchased, it produced a monthly world record of 55,521 tons at its ten-ton capacity per vessel, two-converter Bessemer shop. This was equal to an average of 13,880 tons *per week*, which is ten times more than they were making at the world-record-holding ET plant near the beginning of their operation about twenty-five years earlier, and nearly twice as much as ET was capable of producing in the later 1880s slightly more than ten years before! What makes this even more impressive was that at the same time, the Duquesne Works also produced a world record of 40,321 tons of open hearth steel, passing the Homestead Works.[14] That was a truly phenomenal achievement, surpassing two of the most recognized facilities in the world!

In 1904 *The Iron Age* published a table (table 8.1 shown below) that presented the total rail production by year of Carnegie's Edgar Thomson Works from the commencement of its operations in 1875 up to and including 1903, its third year under the U.S. Steel banner. There was a striking comment made by this world-famous magazine, while reporting mundane statistics: "This table [recreated here as 8.1] presents the most remarkable metallurgical record that has ever been printed."[15]

> Much of the progress of this country in the manufacture of Bessemer steel rails has been due to the enterprise displayed by Andrew Carnegie at the Edgar Thomson Steel Works . . . From year to year Mr. Carnegie steadily increased the capacity of the Edgar Thomson Works and thus cheapened the cost of producing rail . . . he had an unbounded faith in the future of the steel rail . . . He

Table 8.1. Edgar Thomson Steel Production, 1875–1903

Year	*Gross Tons Production*	*Year*	*Gross Tons Production*
1875	5,853	1890	332,942
1876	32,226	1891	264,469
1877	48,826	1892	330,511
1878	64,503	1893	230,336
1879	76,044	1894	220,337
1880	100,095	1895	324,778
1881	151,507	1896	300,776
1882	143,561	1897	477,363
1883	154,892	1898	561,757
1884	144,090	1899	604,343
1885	126,656	1900	626,831
1886	173,001	1901	708,113
1887	192,999	1902	709,906
1888	148,293	1903	675,214
1889	277,401	—	—
Total:			8,207,623

> foresaw this evolution and fully prepared for it when experienced manufacturers and even many railroad officials continued to praise the iron rail. Hence, when others were timid or neglectful of their opportunities, he introduced at the Edgar Thomson Works from time to time the best and most economical methods of manufacture; the blast furnaces at these works were the best in the country, the Bessemer converters were the largest, and the rail mill was the swiftest; so that, when an extraordinary demand for steel rails would come, as it often did come, he was fully prepared to meet it and at a lower cost than that of his competitors . . . The best engineering talent in the country was engaged to bring the Edgar Thomson Works up to the highest possible state of efficiency . . . Mr. Carnegie set the pace for a large annual tonnage of steel rails, and this policy was afterwards applied to the production of pig iron and other products. His American competitors were soon compelled to abandon their conservative ideas and to enlarge the capacity and increase the efficiency of their works.[16]

James Swank, who was not an employee of Carnegie, made these comments. He was a noted author, the long-term (decades) secretary of the American Iron and Steel Association in Philadelphia and one of the leading authorities on the iron and steel industry in the United States, if not that of the entire world. The article was written several years after Carnegie had retired from business.

OPEN HEARTH STEEL

In 1886 when Carnegie began building his first open hearth steelmaking shop at Homestead, there were already seventy-one operating open hearth furnaces listed in the United States, of which approximately thirty-four were located in Pennsylvania with about seventeen of that total in the Pittsburgh Region; that is, Allegheny County. The data is approximate because the number of furnaces at Park was unknown and not included here, although in 1882 their works were known to have two furnaces. That information does not include the Homestead furnaces that were still under construction as of the date of record for the directory.[17] So Carnegie's new facility was not totally unique. As far as American basic steelmaking is concerned, the nation's initial basic Bessemer furnace at Bethlehem was deemed to be inappropriate due to the low levels of phosphorus typically found in American irons—too low to be practical. The first and only basic open hearth furnace built at the Otis Works in Cleveland was determined by them to be too slow, and therefore uneconomical for the market at that time. It was returned to acid steelmaking after operating for only slightly more than two months in early 1886. Later that same year, the Homestead furnaces were started on the acid process, and then afterward were changed to basic steelmaking, becoming recognized as the first commercially successful basic shop in the United States. The Homestead furnaces were similar in design to those at the Park Brother's Black Diamond plant across town from Homestead (on the Allegheny River side). This was not at all surprising because Park was an early advocate of open hearth steel and had an experimental works operating with Siemens's direct process in Pittsburgh as early as 1877. (It later failed.[18]) The furnaces at Homestead were built according to the circular or oval hearth configuration of Lash, a shape initially considered suitable for burning natural gas, an idea that was later discarded. Lash had worked for Park and then for Carnegie, so this also was not a surprise.

As usual, after a brief delay, Carnegie's new steelmaking adventure took off. The original shop, initially planned for four furnaces by the end of 1886, soon had eight, and in rapid succession a second shop with eight more came on line by 1890. This was soon increased to sixteen. In 1897 sixteen furnaces began operating in yet a third shop at Homestead, and this was eventually increased to twenty-four furnaces of forty-five-tons capacity. Open hearth steelmaking was then introduced at the Duquesne Works in 1900 with twelve fifty-ton furnaces set in a single line.[19] In the beginning, the Carnegie Company preferred simple, stationary furnaces at ground level with in-line regenerator chambers, located within each flue tunnel to the stack. This type of construction was relatively inexpensive to build and easier to maintain in comparison to the slightly later, more complicated style of moving furnaces. In the earlier era, they were using hydraulic cranes, and low furnaces facilitated the movement of raw materials and scrap at grade. Additionally, the simple furnaces could be hand charged and/or repaired through doors on both sides of the furnace with relative rapidity.[20]

The pathway to greater production was apparent. With each successive shop there were more furnaces, or furnaces with larger capacity, bringing increasing levels of automation over the previous generation not only in newer shops, but also within existing facilities. Carnegie's open hearth operations were established in fairly quick progression in 1886, 1890, 1897, and 1900. The direction of change and the level of improvements in the Homestead shops should be fairly obvious from an inspection of the plan layouts presented in figures 6.3, 6.4, 6.6, and 6.9 found in chapter 6 on open hearth steelmaking. After establishing itself as the leader of basic steelmaking technology in America at the #1 shop of 1888, in slightly more than two years the Carnegie Company made a prompt move to eliminate the abysmal conditions that were coincident to the use of the central casting pit in the original shop by placing the casting pit at the far end of the new shop, removing it away from interference with the furnaces. Although a step in the right direction, changing the location of the pit proved to be of little benefit to the men who still had to work down in the new pit area, the one exception being that it removed the men from exposure to the tapping and teeming dangers that were often coincidental to those circumstances when they occurred simultaneously at the #1 shop. The original intent of the central pit area was to create the ability to tap two or more of the furnaces sequentially into a single mould to make large ingots for forging and/or rolling into plate. Although the concept appeared to be rational and sound, in actual practice conditions must have been horrific for the workers, as witnessed by the prompt change of the layout for the #2 shop, together with the description found in the *McClure's Magazine* article stating how bad conditions still continued to be at the new-and-improved plant. In the interlude between the inception of the first two and the third shop of 1897, attempts were made to develop improved methodology and devices for mechanically charging scrap, pig, and other raw materials into the furnaces. Finally, a wise decision was made to purchase and install a number of Wellman's proven electrically operated charging machines (the largest concentration in use anywhere). All raw materials were charged cold into the furnaces at this point. By 1894 (but likely years earlier) they had also installed electric overhead traveling cranes in several places throughout the plant in addition to the casting pit, shown in photographs found within the Carnegie Museum of Art. Electric arc lighting was also tried in the shop. The #3 shop of 1897 continued to have furnaces built at grade in two lines just like the first two, but this shop from its inception was constructed with both electric traveling cranes and Wellman charging machines. Casting ingots on cars replaced the abhorrent casting in pits, except very large ingots upward of one hundred tons that required multiple individual furnace heats. By 1900 the Duquesne plant took advantage of relatively new electrically driven material handling equipment and built furnaces about nine feet above grade in a single line, typical of many existing facilities at other companies, and dispensed with the old ladle casting pits. Fifty-ton teeming ladles were also at grade. The earlier furnace ladle pits continued to be a bane to the older number 1, 2, and 3 shops at Homestead, although overhead traveling cranes and casting on cars was eventually adopted in 1 and 2, eliminating these pits. However, ladle pits for each individual furnace were used until the end of their operational lives soon after World War II and the Korean War, when the facilities were finally demolished, having served faithfully for around sixty years.

Meanwhile, Carnegie's competitors were not standing by idly. He was being challenged in the production of open hearth steels in a broad range of markets. Information about some of these businesses is presented in table 8.2 (OH1). Many of the more than one hundred plants in existence by 1902 were small one, two, or three furnace shops, primarily interested in the production of castings or small forgings. These had little or no bearing on the larger primary operations. So only a few of the larger, more pertinent facilities are included.

Table 8.2. (OH1)

Comparison of Stated Nominal Open Hearth[f] Capacity for Published Years 1886–1901. *Carnegie Steel Co. vs. Selected Competitors.*
Compiled from data in the *Directories of Iron & Steel Works—AISA.*[e]

Carnegie Steel Co.			*Penn. Steel Co.*			*Beth. Iron Co.*			*Illinois Steel*			*Cambria Iron Co.*			*Otis Steel*			*Cleve. Roll Mill*			*J&L Steel*[b]		
OH Fces.			*OH Fces.*			*OH Fces.*			*OH Fces.*			*OH Fces.*			*OH Fces.*			*OH Fces.*			*OH Fces*		
Tot. # Fces.	Cap. Tons	Total Tons[g]	Tot. # Fces.	Cap. Tons	Total Tons[g]	Tot. # Fces.	Cap. Tons	Total Tons[g]	Tot. # Fces.	Cap. Tons	Total Tons[g]	Tot. # Fces.	Cap. Tons	Total Tons[g]	Tot. # Fces.	Cap. Tons	Total Tons[g]	Tot. # Fces.	Cap. Tons	Total Tons[g]	Tot. # Fces.	Cap. Tons	Total Tons[g]
1886[c]																							
4*	35[a]	140	2	30	60	2*	15	30	0	0	0	2	15	30	4	15	60	5	7,15	51	0	0	0
1888[c]																							
4	35[a]	140	2	30	60	2*	15	30	0	0	0	2	15	30	4	15	60	5	7,15	51	0	0	0
1890[c]																							
15	15, 20, 35	285	2	30	60	4*	10, 20, 30	90	0	0	0	2	20	40	7	15	105	5	7, 15	51	0	0	0

1892[c]	16	12,20,25,35	367	5	5,15,30	95	4*	10,20,35	95	0	0	0	3	20	60	7	15	105	4	7,15	44	0	0	0
1894[c]	16	12,20,25,35	367	12	5,7,15,30,50	402	4	10,20,40	110	4*	15	60	3	20	60	7	15	105	2	15	30	0	0	0
1896[c]	20	12,20,25,35	507	12	5,7,15,30,50	402	6*	10,20,40	190	10*	20,30,50	340	3	20	60	7	15	105	2	15	30	0	0	0
1898[c]	15	12,25,35,40	1042	12	6,20,40,50	402	7	10,20,40	230	10	25,50	350	4	20	80	7	10,18	118	2	15	30	7	25,40	265

(continued)

Table 8.2. *Continued*

Carnegie Steel Co.			*Penn. Steel Co.*			*Beth. Iron Co.*			*Illinois Steel*			*Cambria Iron Co.*			*Otis Steel*			*Cleve. Roll Mill*			*J&L Steel*[b]		
OH Fces.			*OH Fces.*			*OH Fces.*			*OH Fces.*			*OH Fces.*			*OH Fces.*			*OH Fces.*			*OH Fces*		
Tot. # Fces.	Cap. Tons	Total Tons[g]	Tot. # Fces.	Cap. Tons	Total Tons[g]	Tot. # Fces.	Cap. Tons	Total Tons[g]	Tot. # Fces.	Cap. Tons	Total Tons[g]	Tot. # Fces.	Cap. Tons	Total Tons[g]	Tot. # Fces.	Cap. Tons	Total Tons[g]	Tot. # Fces.	Cap. Tons	Total Tons[g]	Tot. # Fces.	Cap. Tons	Total Tons[g]
1900																							
Only a supplement to the 1898 directory was published, containing information on the large number of consolidations rather than updated furnace data.																							
1902[d]																							
56	20,40,45,50	2460	14	6,20,40,50	472	8	10,20,40	270	10	30,35,50	385	10	20,50	380	10	10,18,25	178	4	50	200	9	15,25,40	280
Carnegie vs. All Others listed in this table for 1902																							
Carnegie's total was fifty-six furnaces versus seventy-five furnaces for all others listed.											Carnegie's stated total furnace capacity of 2,460 tons exceed the sum total of 2,165 tons for all others in this table.												

*Some or all of the new furnaces were stated as being under construction during year.

a—Stated capacity of thirty-five tons for the 1886 and 1888 time frames is not supported by facts in other documents. Original total capacity was more near sixty tons or less and not 140.
b—The J&L furnaces listed for 1902 includes two from the Soho Iron (Steel) Company, which were purchased by them. The Soho plant, of which the two open hearth furnaces were a part, was built prior to 1886. Soho was also one of the older blast furnace plants in Pittsburgh and dated to 1872, the same year that Lucy and Isabella were built.
c—Some issues of the directories reflect conditions as of the previous year; others are during the given year. For instance: 1886 is as of July 1886; 1888 until November 1887; 1890 to November 1889; 1892 to February 1892; 1894 to March 1894; 1896 to January 1, 1896; 1898 to April 1898.
d—1902 information is from the 1901 *Directory* (15th edition) which includes data correct to December 31, 1901, and is therefore current for 1902.
e—American Iron and Steel Association.
f—Data is indicative of the number of furnaces whether acid, basic, or a combination of the two.
g—This figure had been computed by multiplying each nominal furnace capacity (second column) by the number of furnaces with that stated capacity (not included in this table but found in each *Directory*) and summing for a total for each company for the given year.

An approximation of actual daily maximum production can be computed by considering that the cycle time for one heat in an open hearth furnace was about ten to twelve hours at this time. Multiplying the number in the total tons column by a factor between 2 and 2.4 (estimated number of heats for twenty-four hours) will give an estimate of a company's potential output in tons per day. The table does not reflect that, however; it can neither account for lost production due to problems or repairs nor low levels of utilization due to economic downturns, work stoppages, shortages of raw materials, or lack of orders. Some plants made steel with tilting furnaces, through "duplexing" by the Talbot process or modifications thereof (both explained a bit later), and therefore they could produce open hearth heats somewhat faster than single or stationary furnaces. However, other factors affecting conversion rates make it difficult to estimate their cycle time.

Regardless, a good understanding of the Carnegie Company's status in the American open hearth steel market can be readily gained. Ultimately, the most stunning detail that arises from the information is that by the end of 1901, the Carnegie Company, with fifty-six furnaces and 2,460 tons of stated furnace capacity, was capable of outproducing the sum total of the rest of its major competitors, together with a few minor ones added in as well, that are included in the table. The competitors' seventy-five furnaces were capable of only 2,165 tons of capacity (see chart 8.1). As reported from London in early 1902, the Carnegie Homestead Works was the largest and the most important open hearth shop in the world.[21]

It should be noted that between 1896 and 1902 there was literally an explosion in the number of open hearth plants in the United States. For example, large-capacity facilities in Clairton, Pennsylvania, with twelve fifty-ton furnaces and the soon-to-be-built plant at Donora, Pennsylvania, with an identical number and size of equipment, were not even on the radar at the beginning of 1896 (a financial depression year). There were also some other large facilities not included in the table. A few that are worthy of mention are Pencoyd, near Philadelphia (one seventy-five ton and ten thirty-ton open hearths); Sharon Steel (Sharon, Pennsylvania) with eight fifty-ton existing furnaces and five being built; and Sharon (Pennsylvania) Works of National Steel with six thirty-ton furnaces. These were typically plants that specialized; Pencoyd was a structural maker, for example. The United States surpassed England in open hearth steel production during this era (1900) to become the world's largest steelmaking nation for both Bessemer and open hearth steels. In 1901 American open hearth output almost equaled that of Great Britain and Germany combined.[22] Open hearth steel production had been on the rise and was beginning to take the premier position in the American market, but it wouldn't surpass Bessemer output until 1908.[23]

In only one issue of the *Directory* covered by the aforementioned table (1894) did any company exceed Carnegie's total tonnage of stated capacity. That was accomplished by the Pennsylvania Steel Company with 402 tons versus Carnegie's 367. It also appears that others probably exceeded Carnegie's *actual* tonnage in the beginning (1886 and possibly 1888). Otis had four fifteen-ton furnaces, and Carnegie didn't build the thirty-five-ton furnaces until later than reported. Originally Carnegie constructed smaller furnaces of fifteen tons. A similar argument might be made for Park Brothers at their Black Diamond Steel Works (not included in the table), but it is difficult to find data for this company. *American Manufacturer and Iron World* for 1887 stated brick was being laid for a fifth thirty-ton furnace at the Park Works that had two other fifteen-ton furnaces already in existence. This far surpassed the abilities of the Homestead Works.[24] However, after Pennsylvania Steel's lead reported in 1894, data in the report for 1896 clearly shows that Carnegie was back on top with 507 versus 402 tons, and according to the 1901 volume (1902 data line), Carnegie's lead had widened and had exceeded Pennsylvania Steel's abilities (then second in order of ability) by more than 500 percent! This despite the fact that Pennsylvania Steel was blessed with an innovative manager, H. H. Campbell, who was a respected steelmaking authority and author in general and about the open hearth process in particular.

Starting with two fifteen-ton furnaces in 1875, the Steelton plant had an early history in the operation of open hearth furnaces in America. This predated the Homestead facility by eleven years.[25] Campbell developed the first open hearth furnace with a tilting design in order to accelerate open hearth steelmaking, among other things. Shortly afterward, Wellman, the inventor of the first practical open hearth charging machine, introduced his own tilting mechanism for an open hearth. Campbell's furnace rotated on rollers contained within a curved track, while Wellman's pivoted around a hinge point. Both used large hydraulic cylinders to power their movements.[26] Steelton also implemented something known as the "duplex process," which in this particular instance refers to the use of a Bessemer converter in tandem with an open hearth furnace in sequence to remove rapidly most of the carbon and silicon from the bath in the converter and then finish the heat in the open hearth to attain an accurate chemical endpoint. This significantly reduced processing times. There were other duplex processes. The term simply meant the use of two furnaces or vessels to complete the operation, such as in Bertrand-Thiel, with the most popular application of the term referring to the Bessemer–open hearth combination that continued to be used in the United States until about 1970.

Comparison of Homestead Works Facilities 1890 versus 1898
Table from *Trust Finance*, Edward Meade[27]

1890

1. Two 5-ton Bessemer Converters.
2. Seven open hearth furnaces—one 15-ton, four 20-ton, two 35-ton.
3. One 28-inch blooming mill.
4. One 23-inch and one 33-inch train for structural shapes.
5. One 10-inch mill.
6. One 32-inch slabbing-mill for rolling heavy ingots.
7. One 120-inch plate-mill.

Annual capacity, 295,000 tons.*

1898

1. Two 10-ton Bessemer converters, one 12-ton.
2. Thirty open hearth furnaces—one 12 ton, six 25-ton, eight 35-ton, and fifteen 40-ton.
3. One 28-inch and one 38-inch blooming-mill.
4. One 23-inch and one 33-inch train for structural shapes.
5. One 10-inch mill.
6. One 32-inch slabbing-mill.
7. One 40-inch cogging-mill.
8. One 35-inch beam-mill.
9. One 119-inch plate-mill.
10. One 3,000-ton and one 10,000-ton hydraulic press.
11. Steel foundry, press shop, and machine shop.

Annual capacity, 2,260,000 tons.*

*Note: This is an 870 percent increase in production capacity in only eight years, during which, it should be mentioned, a severe financial depression occurred in the United States during the 1893 to 1896 period. This also represents an almost unbelievable 3,700 percent increase in capacity over the plant's original design of sixty thousand tons per year that was accomplished with two 4-ton converters, a billet mill, and a rail mill in 1881.[28]

Chart 8.1. Open Heath Furnace tonnage available to Carnegie and his chief competitors nationally and locally. Compiled from data in table 8.2 found in the *Directory of Iron and Steel Works of the United States and Canada*, 15th edition (accurate to 12/31/1901). Figures reflect the sum of the stated tonnages of furnaces times the number of furnaces in the plant at that rated tonnage. Note that by this time, at Homestead and Duquesne Carnegie maintained in excess of five times the rated capacity of his next nearest competitor. Carnegie, Illinois (Federal) and the Cleveland Rolling Mill would be merged into U.S. Steel; Pennsylvania, Bethlehem and Cambria became part of the new Bethlehem Steel Company under Schwab and some four decades later, Otis would become a part of J&L. Carnegie's production was from basic lined furnaces, other firms could be either acid, basic or a combination of both.

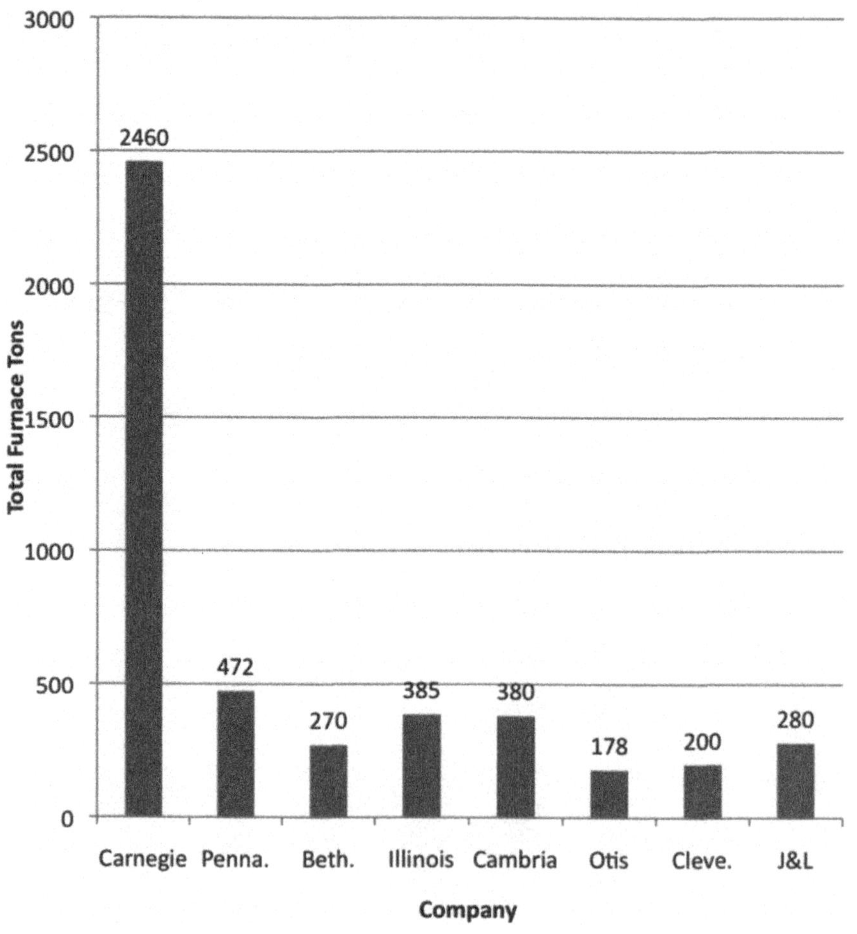

According to a report by the British Iron Trades Association in 1902:

> The development at Pittsburgh alone represents an enormous increase in producing capacity, and the latest additions at Duquesne and Homestead are brilliant examples of modern melting shops, which, for boldness of conception, efficiency of design, and excellence of erection, have seldom, if ever, been known in the history of open hearth steel making.[29]

RIVALS, CONTENDERS, AND CANDIDATES

The Illinois Steel Company, a major national adversary that was a serious opponent of Carnegie in ironmaking and Bessemer in the steelmaking area, did not venture into open hearth steelmaking until late 1894. When Potter and Forsyth were in charge, they challenged Carnegie and sometimes jumped ahead in either ironmaking or Bessemer.[30] In 1889, the newly formed Illinois Steel Company was touted as being the largest iron and steel corporation in the world.[31] Because it did not enter the open hearth business until 1894, and then only with a small four-furnace shop, it was far behind. By 1896 there was only a smattering of ten furnaces of different sizes at South Chicago (which became U.S. Steel's famous South Works). The new steelmaking facility consisted of furnaces with both acid and basic linings, and of both stationary and tilting structures. In 1902 there was no change in the number of furnaces and little change in capacity.

The venerable B. F. Jones (not Captain Bill) led the firm of Jones and Laughlin and its American Iron Works located on the south side of Pittsburgh. J&L was an early entrant into the manufacture of pig iron and wrought iron in the city, but it was a latecomer to the steel trade. From their origins in 1853, about the time a young Carnegie was only a telegrapher, it became one of the leading wrought iron firms in the region. In 1859 the company originally planned to build the first modern blast furnace in Pittsburgh ahead of Graff Bennett's trailblazing Clinton Furnace, but they held off until 1861 due to a business recession. Perhaps the fundamental understanding and/or the recognition of the extreme importance of Connellsville coke might not have occurred at all or could have been delayed or even lost if J&L's Eliza Furnaces rather than Clinton had been the first to operate. Some experiments were tried with skip loading of blast furnaces at Eliza in the early 1880s, but photographic evidence shows the efforts appear far from the mark. J&L was a rather late entrant in Bessemer steelmaking, not beginning until 1886.[32] There are hints that it might have been tried earlier, even before Carnegie, but then withdrew, although it is not at all certain.[33] J&L was a pioneer in the development and production of cold-finished bar stock, a very significant trade. Their own natural gas wells were drilled in 1884. These wells were located inside the plant within the city limits of Pittsburgh. Open hearth production began in the late 1890s, but puddling furnaces were maintained until nearly the end of the century. In the early 1900s, J&L built an enormous tilting furnace of about two-hundred-tons capacity for making steel rapidly by the Talbot process, and although this began a new era of massive-sized furnaces, Talbot steelmaking never really worked satisfactorily.[34] In 1916 the first "torpedo" or mixer-type hot metal transfer or ladle car was invented that had a ninety-ton capacity. The car was devised to reduce the heat losses involved in the movement of molten iron from J&L's Eliza Furnaces to the mixers at the open hearth or Bessemer shops across the river or between Pittsburgh and its new plant in Aliquippa.[35]

J&L, a well-respected firm, was never a significant steelmaking threat to Carnegie on any plane. It did make one other highly significant, although totally unrecognized, stride in the area of coke making. It brought coal for coking from the mines to the plant, instead of making coke at or near the mine and then having it shipped to the premises. This practice was not embraced by other firms for a decade or two, when there was a shift away from beehive coking to the newer by-product slot oven technology. Had other companies recognized this potential earlier, they could have shipped the prized coal from the Connellsville region to their own plants, Chicago, for example, and made the coke there without having to account for the high percentage of breakage that occurs with long rail shipment and the necessary handling. Coal is not fragile; coke is. But as it takes 30 to 35 percent more coal by weight to make coke, shipping coal by rail would have been considerably more expensive because of that fact. This,would have been offset somewhat by coal's higher density and lower volume, but that could be overcome through taking advantage of very low shipping rates via a barge on the river, as J&L did, and then by sending it to some intermediate river port much closer to Chicago (for example) in addition. Then, there would be a much shorter and therefore less expensive trek by rail to the plant and less breakage, too, once the coke was made there. Shipping coke by barge was not practical due to even higher losses from breakage incurred when compared to shipping by rail. Coal was at a much lower selling price and considerably easier to ship as well, but the conversion cost would then have been borne by the receiving steel company.

Both the Cleveland Rolling Mills and Otis began employing open hearth steelmaking technology very early compared to Carnegie, actually at about the same time that Carnegie began Bessemer production. Otis was the first all open hearth steel producer on a large scale with first two and then four furnaces. It was also the first in this country to use a basic open hearth, although not for long. Even with this extensive lead in understanding the process, according to the figures presented in the table, neither Cleveland nor Otis ever really mounted any serious competition once Carnegie got established. This fact possibly reinforces the important relationship between open hearth production and the use of in-house Bessemer scrap. Samuel Wellman, the inventor of the first practical charging machine, worked for Otis and left shortly after that development. That was perhaps one reason for their stagnation.

The Bethlehem Iron Company was a special case. Bethlehem was Carnegie's chief (only) rival in the armor plate market. They were the only two companies producing specially treated alloy steels in the appropriate quantities and required quality demanded for America's capital shipbuilding projects. All others were only very minor participants in that market. Carnegie, a proclaimed pacifist, refused to make forgings for guns. However, despite that fact he somehow convinced himself that it was okay to make armor plate for the ships themselves. Somewhat later though, he had to be restrained by his partners from entering the armament markets. Such was the nature of his personality. The Bethlehem Company was already producing Bessemer steel a few years prior to the Edgar Thomson Steel Company's entry into the business. Recall that it was also the first basic steel producer in the United States (albeit in a Bessemer converter), but it weaned itself from that process as it began open hearth operations in the early 1890s. Bethlehem began to rely solely on open hearth production after acquiring a selection of some acid and some basic furnaces near the turn of the century. It concentrated on a very limited market, producing armor and regular plate, large forgings, and castings. This company's name was the standard bearer of a new conglomerate, as the Bethlehem Steel Company, when Charles Schwab, the last chief executive of Carnegie Steel, formed the new company after he left the presidency of U.S. Steel under less-than-friendly circumstances (see table 8.3 for some of the merging firms).

Cambria, like the Pennsylvania and Illinois (South Chicago) Steel Companies, was a major national firm far in advance of Carnegie entering the steel business. Cambria initially delved into a more sophisticated practice of open hearth steel making using a Krupp washer, but as can seen in table 8.3, by 1902, with the exception of Carnegie, it was in roughly the same ballpark of scale as the majority of competing firms found in table 8.2 (OH1).

OTHER FACTORS

> The policy of the Carnegie Company was purely industrial. Financial considerations had little weight. Its shares were never in the market. The greater part of its profits was each year invested in the plant. As Mr. Carnegie remarked, he and his partners knew little about the manufacture of stocks and bonds. They were only conversant in the manufacture of steel.[36]

Table 8.3. Comparison of Some Major American Steel Companies Post-Carnegie Era as of 1904

Company Name	Works or Location	# Blast Fces. (coke)	Bessemer Converters (# vs. size t.)	Open Hearth Furnaces Acid	Open Hearth Furnaces Basic	Open Hearth Furnaces Tot.	Comments
Carnegie Steel Co.[c]	Edgar Thomson	11	4 × 15	0	0	0	Produced rails, plate, armor plate, bar, structural steels such as beams and angles, and also large forgings.
	Homestead	6	2 × 10	0	50	50	
	Duquesne	4	2 × 10	0	14	14	
	Lucy	2	0	0	0	0	
	Total	23	8	0	64	64	
Federal/Illinois Steel Co.[c]	South Chic.	10	3 × 15	0	10	10	Primarily produced rails and related items.
	Joliet	4	2 × 10	0	0	0	
	Union	2	0	0	0	0	
	North Chic.	2	0	0	0	0	
	Milwaukee	1	0	0	0	0	
	Total	19	5	0	10	10	
Bethlehem Steel Co.[a]	Bethlehem PA	4	4 × 7.5[b]	7	4	11	Plate, armor plate, large forgings and castings.
Cambria Steel Co.[a]	Johnstown PA	6	4 × 12.5	2	12	14	Rail, plate, structural steel, bars.
Maryland Steel Co.[a]	Sparrows Point MD	4	3 × 18	0	0	0	Primarily rails.
Penna. Steel Co.[a]	Harrisburg/ Steelton PA	7[d]	3 × 10	5	9	14	Primarily rails, also structural and plates.
Lackawanna Steel Co.[a]	Lackawanna NY	3	4 × 10	0	0	0	New in 1901–1903, six open hearths u.c.
Jones & Laughlin Steel Co.	Pittsburgh	5	3 × 10	1	7[e]	8	Structural steels, bars, plates. Cold finished bar.
	Soho (Pitts.)	1	0	4	0	4	
Otis Steel Co.	Cleveland	0	0	2	8	10	Plate, bars, forgings, castings.
Clairton[c]	Clairton PA	3	0	1	11	12	New 1901–1902
Donora[c]	Donora PA	2	0	0	12	12	Blast Furnaces not in operation 1904.
Pencoyd[c]	Philadelphia	0	0	0	11	11	Structural related to bridge building.
United States	**1904 totals**	287	75	185	346	549	# Plants 1904/1901 Bess.-32/35, OH-135/112
	1901 totals	NA	81	147	236	403	

a — Soon became part of the new Bethlehem Steel Company under Charles Schwab.
b — Abandoned, demolished in 1902.
c — Became part of U.S. Steel Company in 1901.
d — Used a combination of anthracite and coke.
e — One furnace had 250 tons capacity. Normal in the United States approximately 30–50 tons.

Source: Table compiled from data contained within the *Directory of Iron and Steel Works of the United States*, sixteenth edition (Philadelphia: American Iron and Steel Association, 1904).

Kloman, Jones, Frick, and Schwab were probably the most notable assets that Carnegie had the luck to acquire as associates. As previously discussed, Kloman became a liability due to his poor personal business practices that brought the entire company to peril, and so he was eliminated. Jones, while a definite asset, died before his time in a mill accident, and he was the only member of this elite group never to become a partner. He guided a young Charlie Schwab in the intricacies of mill operations and management at ET, although Schwab was a very able manager in his own right. In 1886 Jones recommended that Carnegie appoint him to the position of assistant superintendent under Julian Kennedy at the Homestead plant. Kennedy unexpectedly resigned in 1888, and Schwab was there ready and able to assume the vacant spot. When Jones died suddenly at the end 1889, Schwab was recalled to direct the flagship at Edgar Thomson. He was now a much more highly regarded resource, as Schwab understood the ramifications of both the Bessemer and open hearth processes in detail since he had improved both costs and production at Homestead. At ET he further improved upon Jones's administration plant output and profits. Part of the reduction in costs came from Schwab's knack for pitting one group of employees against another. Jones also had used competition, but in a much less controversial manner. Schwab also acquired an ability for dealing with strikers, a competency that would soon come in handy in 1892. Carnegie began to believe that Schwab was at least the equal of Jones. Schwab took over control of Homestead again in the aftermath of the tragic 1892 strike.[37] Frick and Carnegie needed him to unruffle the feathers of the returning workers, those who were not blacklisted.

Frick was one of the ablest managers of his time. He did not like Carnegie's interference. Carnegie had made a serious error in not buying out the H. C. Frick Coke Company in its totality. That would have been difficult to accomplish, but not for financial reasons, as Frick might not have agreed to let that happen. Each man continued to maintain a certain soft spot for his own creation—Carnegie for steel and Frick for coke.

Phipps demanded that the company adopt a policy to protect the shareholders, especially in the event of Carnegie's death. In the tragic year 1886, the year that Carnegie's mother and brother died, events nearly brought Phipps's fear to fruition since Andy was near death at that same time. Consequently, the company adopted a policy, known as the "iron-clad agreement," wherein a partner's share of the business could be bought out under certain conditions. The principal intention was to be able to buy Carnegie's share at book value over time if he died without destroying the company in the process, since he owned more than 50 percent.[38] Unfortunately, the agreement was used as a tool to eliminate certain partners who had been granted a tiny share of the business as a reward for excellent service. Those who fell from grace were removed at book value. Near the end of its tenure, the book value of the Carnegie Steel Company was $25 million. The true value of the company at its sale was nearly $480 million.

In 1899 the Carnegie Steel Company had become embroiled in a dispute that erupted into a major crisis because of an argument between Frick and Carnegie. The disagreement involved the price of coke sold to the steel company by the coke company. Frick and Carnegie had settled on a long-term price for coke for the steel company. Neither man should have negotiated such a deal since neither was an executive for either company. One was the chairman of both companies, and the other the principal shareholder of both. Ultimately the price was not met due to a series of circumstances, resulting in the argument between the two, with Frick being ejected from the Carnegie Steel Company via the iron-clad agreement. Frick owned about 6 percent of the shares. Consider the enormous difference in the value of those shares measured at the $25 million capitalized value versus the $250 million estimated value for 1899 or by the actual sale value of $480 million to Morgan in 1901.[39]

The root of this turmoil between Carnegie and Frick can be traced back to Carnegie's loss of stature and esteem brought about by the 1892 Homestead Strike, as well as Frick's unsuccessful effort to entice Carnegie to sell the company to W. H. Moore and J. W. (Bet-a-Million) Gates in 1899. Frick secretly had provided funding to them to help move it forward, and Carnegie discovered the misdeed. All of this finally reached the boiling point over the price of coke in 1899.

Again, Schwab was waiting in the wings to take over the chairmanship of the company. It was Charlie Schwab who finally convinced Carnegie to sell the company to Morgan in 1901, after Schwab was motivated by Louise Carnegie to do so. Schwab became the first president of the newly formed United States Steel Company, but he lost the chairmanship to Morgan's partner, Elbert Gary. With his loss to Gary, Schwab left the company, soon becoming the U.S. Steel Company's nemesis by forming the Bethlehem Steel Company through merging Pennsylvania Steel, Cambria Steel, Maryland Steel, Bethlehem Steel, and Lackawanna Steel. Bethlehem became this country's second largest steel producer for most of the twentieth century.

For a greater understanding of the complicated story and the people involved in this controversial period of the Carnegie Company, see *Industrial Genius* (about Schwab) and *Triumphant Capitalism* (about Frick), both by Kenneth Warren, and *Andrew Carnegie* by Joseph F. Wall.

Carnegie was, more or less, the pioneer of the basic steel industry in the United States. So it was his company and his company alone that could initially take advantage of the large volumes of phosphorus-containing scrap available at that time, especially since it was one of the larger producers of that kind of scrap in America and had access to that portion of it for no additional cost. This put Carnegie at a serious competitive advantage over the other open hearth producers since open hearth steel was considered to be superior in the mass-produced steel market. Acid producers were limited to using only the more difficult-to-find and therefore more expensive, higher-grades of lesser-contaminated kinds of scrap. Bessemer producers that chose to use home scrap were limited by the volume of scrap that could be consumed as well as by the productivity decrease coincident to adopting the policy of scrapping. Through Carnegie's realization of a commercial-quantity basic plant, Bessemer scrap transformed from a material with limited utility into a desirable commodity that could be used for conversion into more valuable basic steels. Acid open hearth operators were also disadvantaged by the general use of producer gas for fuel, which was made from coal. Contact between the flame and the metal bath transferred some of the sulfur from the flame into the metal, joining with any phosphorus present to further contaminate the steel. Acid furnaces could remove neither. Recall that sulfur is responsible for "hot" or "red" shortness and phosphorus for "cold" shortness in the final product if the level of either becomes too high. Carnegie, through his wise choice of natural gas, a relatively abundant, inexpensive, readily available, low-sulfur fuel that was easily found in the Pittsburgh region at that time, was not plagued by these problems. He invested in land containing productive gas fields, drilled the wells, and assembled the infrastructure to distribute the fuel to his plants. He wasn't bothered by the negative costs or complications commensurate with generating producer gas from coal.

Carnegie also introduced charging molten iron into open hearth furnaces, probably as a means of speeding up the cycle time, saving fuel, and increasing productivity.[40] He could only do this to a limited extent at Homestead because all of the iron delivered in the molten state had to come from either Duquesne or ET. This practice was restrained by the volume available and was restricted to the excess supply over and above the metal needed for use at the converters at those facilities. Homestead did not have a blast furnace plant of its own until Carnegie purchased the Carrie Furnaces (Pittsburgh Furnace Company) in Rankin in 1898. Even then it had to wait until they built a bridge between the plants before it could be brought directly across the river as a liquid. Charging molten metal was not new. It was used as early as the days of Martin's and Siemens's experiments as well as in numerous other times without much success.[41] Molten iron additions proved to be quite corrosive to the furnace refractory, depending on the means by which they were accomplished.[42] Apparently the success of the basic open hearth process caused a shortage of scrap in the industry, and there are indications that scrap became both more scarce and expensive.[43] Beginning in early 1900, Carnegie adopted a method of making open hearth steel using the Monell process in two furnaces at Homestead, at least on a test basis. Monell was a Carnegie employee. He devised a modification of the pig and ore (Siemens) process, where he charged molten pig iron directly onto a bed of highly heated iron ore and limestone that was charged cold some time prior to introducing the pig. There was a large and immediate reaction when the newly charged molten pig was fed into the furnace and steel was made very rapidly in seven or eight hours as opposed to the ten to twelve that was typical. No scrap was needed.[44] The application does not seem significant, possibly due to Carnegie's exit from the industry shortly after the beginning of the trial. Following a brief spurt of interest in the United States, this and other variations of the pig and ore process (another form being Talbot), which claimed large increases in metallic yields, soon fell from favor.

A major unseen outcome that emerged from the plethora of developments undertaken by Carnegie, a catalyst motivating the integration of many competitor's steel plants, was that a modern, high-production, high-efficiency steel mill had to imitate Carnegie's in order to be efficient. With the introduction of the Jones mixer, the blast furnace and the Bessemer plant had to be located in close proximity to each other, with the mixer near the converters. When the basic open hearth shop became a consumer of Bessemer and other in-house scrap, especially after the routine introduction of molten metal charges began, locating the open hearth plant near the blast furnace and the Bessemer department became more of a necessity because the open hearth had to use its own mixer for molten metal or share one with the converter shop. Having rolling mills near or adjacent to the furnaces, while not critical, was only logical. So in order to be efficient, a modern high-production steel mill had to be configured as a two-bell, automatic, electrically driven bucket or skip-fed blast furnace plant with pig casters, and have all of the proper appurtenances found in an ore field such as ore bridges, storage bins, scale cars, and others. These facilities needed to be located very close to a mixer and converter shop and/or a basic open hearth shop and mixer, with all adjacent to the rolling mills. That was the exact layout at Edgar Thomson, Homestead, and Duquesne, with the unintended result from the efforts of the Carnegie Steel Company. Homestead did not follow this pattern until after the purchase of the Carrie Furnaces, shortly after the introduction of hot metal charges. Prior to that it used Bessemer and other scrap from all three plants, along with the cold pig.

This general practice was not superseded until the late twentieth century with the introduction of the basic oxygen furnace (BOF), in essence a modified Bessemer converter that is blown with pure oxygen through the top instead of the bottom, followed a bit later by the addition of the continuous caster, which replaced the entire process of teeming, stripping, and blooming ingots. Now an efficient mill making steel from ore consists of a blast furnace next to a BOF next to a continuous caster next to a rolling mill. Sometimes a ladle metallurgy furnace and/or vacuum degasser is interspersed between the BOF and caster to trim the chemistry of the heat or to adjust the temperature. In today's world the determination and control of the composition and chemistry of the steel is light-years beyond anything that Carnegie could ever have imagined.

The Carnegie Steel Company was one of the most vertically integrated companies in existence. It was a leader in recycling as well as economizing the use of fuel and many other materials that otherwise would have been wasted. Carnegie Steel would probably receive awards for these efficiencies today. They were constantly getting more from less while simultaneously producing goods of higher quality and value in addition to gaining a greater return on investment.

Yes, Carnegie did cut men's wages, as did his competitors. In Pittsburgh, the decrease in labor cost was often achieved by eliminating men's jobs altogether through process changes and/or automation. A few obvious examples are:

- the pig bed men and cupola men, accomplished through the invention of the mixer;
- bottom filler and top fillers by the use of skip loading and bucket loading;
- material handling of ore and coal with Hulett unloaders, ore bridges, bottom-dump, large-capacity railroad cars and elevated stock bins;
- Wellman charging machines and rail-delivered charging boxes at the open hearth;
- hydraulically operated lift tables on rolling mills;
- casting ingots on cars and using hydraulic ingot strippers;
- electric overhead traveling cranes;
- adopting the use of electric motors that were being developed at the nearby Westinghouse factory, largely eliminating the need for steam-powered equipment and the many men usually required to operate that type of machinery[45]

These and other improvements dispensed with hundreds and hundreds of jobs and their wages, usually for work that was considered to be the most dangerous and/or those more physically demanding.

The scale of the Carnegie Company's reinvestment of profits in some of the later major projects (either in size or for impact) in a relatively short time is astounding when you consider them together:

- 1896 to 1897 construction of the largest blast furnace plant in the world at Duquesne
- construction of the largest open hearth shop in the world at Homestead in 1897 to 1898 that started partial operation only seven months after the beginning of construction
- development and installation of Uehling pig casting machines at Lucy, Edgar Thomson, and Duquesne in 1896 to 1897
- reconstruction of the Bessemer and Lake Erie Railroad in 1897
- construction of the twelve-furnace open hearth shop at Duquesne in 1899 to 1900 that reached world-record-holder status for open hearth output only one year later in October 1901
- development, shakedown, and installation of Hulett unloaders at the Conneaut docks in 1899 to 1900
- purchase of the Carrie Furnace Company in 1898, where expansion of the plant from two to four furnaces began almost immediately after its purchase (the two new blast furnaces built were the largest in size in the world[46]) together with the construction of a railroad bridge across the Monongahela River (the largest/heaviest in the world of its type[47]) to connect Carrie to the Homestead Works in 1898 to 1901 (The main span of the bridge over the river was built in 1900; taking only a few months to complete once the piers were finished.)

Although the scenario for success presented in this summation is complex, it ultimately serves to show that the Carnegie Company's rise to dominance was not accomplished by chance or through chicanery. The inventive and/or managerial genius of Jones, Kennedy, Schwab, Shinn, Neeland, Scott, Kloman, Phipps, Frick, Holley, Brother Tom, Andrew, and numerous others were paramount to that achievement. But in order to make a complete account we must also look beyond the scope of the Carnegie Company and give credit to those who forged the early wrought iron and crucible steel industries in Pittsburgh—the early blast furnace operators; the men at Clinton, who realized the amazing value of Connellsville coke; those that developed the mines and the ovens; and the men of the Monongahela Navigation Company, who refused to take no for an answer and built the locks and dams and transportation network on the rivers without the aid of the government. We must also consider the puddlers, melters, and even the laboring men who gained the skill, knowledge, strength, and stamina necessary to do the jobs that made the larger picture possible. Most were hard working and industrious.

Carnegie, when not in front of everyone else, was always willing to adopt leading-edge technologies introduced by others to get ahead, when either he saw the need or it simply met his fancy. For the most part, he was the aiming point for the rest of the industry. A company can catch up, but at what cost?

Especially when others seemed to be more interested in earning money. With Carnegie, while money was primary, there often appears to have been a rather large, overriding desire to feel important or to soothe his ego. An idea of the power he held is exhibited below.

Following the collapse of the rail pool at the end of 1896 (near the end of a financial depression in America), the Carnegie Company intentionally reduced the average selling price of rails from $28 to $18 per ton in order to capture more of the market. Some sales of rail were made as low as $14.[48] The company claimed that it could operate at a profit even at this lowest sales price, since their cost of production was said to be only $12 at that time,[49] although, based on my recent discussion, this figure probably did not account for any contribution to overhead, if it was in fact possible at all.[50] Profits at Carnegie for 1897 were $7 million (see table 8.4). The great Illinois Steel Company in South Chicago could not compete. It was hurriedly instituting austerity measures to pare costs in order to attempt to meet the realities of the new market. The Carnegie Company could manipulate this market, not only because it was a very cost-effective producer but also it could accurately gage the cost of producing rails, while obviously others, even those who were considered to be the higher-echelon operations, either could not, did not, or were otherwise more focused on trying to know their profits instead. The ability of Carnegie to meet and/or to force others to meet the demands of the expanding rail market, while continuing to make a profit during an almost continual decline in selling price of rails, was amazing. This extended back to the startup of the Edgar Thomson Works (see tables 5.3, 5.4, and 5.5), and undoubtedly it was one key factor in sustain-

Table 8.4. Carnegie Steel Company Profits for Selected Years

Year	*Profits USD*	*Year*	*Profits USD*
1881	2,000,377.42	1891	4,300,000
1882	2,128,422.91	1892	4,000,000
1883	1,019,233.04	1893	3,000,000
1884	1,301,180.28	1894	4,000,000
1885	1,191,933.54	1895	5,000,000
1886	2,925,350.08	1896	6,000,000
1887	3,441,887.29	1897	7,000,000
1888	1,941,555.44	1898	11,500,000
1889	3,540,000*	1899	21,000,000
1890	5,350,000	1900	40,000,000

*Accounting switch in method of reporting profits.

Source: James Howard Bridge, *The Inside History of the Carnegie Steel Company* (New York: The Aldine Book Company, 1903), 102, 295. The table was assembled from profit data found on both of these pages.

ing the amazing period of growth and expansion that occurred in America during the last quarter of the nineteenth century. It was during this time that this nation succeeded in making an overwhelming transformation from an agrarian backwater into an industrial Goliath.

An indication of the impact that the Carnegie Company made on America and its industry, prior to its purchase by Morgan et al. and its subsequent assimilation into the newly formed United States Steel Company in 1901, was noted in a study by a federal commission on industry trusts and combinations that reported it thusly:

> The Carnegie Company manufactured perhaps a larger general variety of steel articles than almost any other manufacturing concern. It produced from 20 to 30 per cent of the steel made in the United States. Of structural materials, plates, etc., it made 50 per cent; of rails, 30 per cent; of armor, 50 per cent. Its exports were 70 per cent of the steel exports of the United States.[51]

The source of the Carnegie Steel Company's total output of raw steel production was limited to the Pittsburgh area alone and came from only three sites in Homestead, Braddock, and Duquesne. These three facilities occupied less than two square miles and could be seen from one to the next to the next. On the other hand, U.S. Steel was an enormous conglomeration formed to stifle competition, and in the process it reduced innovation and increased prices. The results of these new "ideals" from the "Steel Trust" can be inferred by the dramatic drop in market share from the inception of U.S. Steel over a similar time frame (about thirty years) when compared to Carnegie Steel Company's rise to greatness (table 8.5 and chart 8.2). The new "Billion Dollar" steel company's output dropped from a 65 percent share of the nation's steel ingot production in its formation year 1901 down to 41 percent by 1930. The 41 percent share in 1930 is not the result of the Great Depression, since their loss of market showed an almost constant downward trend over the entire thirty-year span of the data (refer to chart 8.2).

Carnegie had been an advocate of the ideals of British philosopher Herbert Spencer, a proponent of something known as "Social Darwinism." It was Spencer who coined the phrase *survival of the fittest* to describe Charles Darwin's "natural selection." Carnegie, who was extremely proud of his accomplishments to date, convinced Spencer to visit Pittsburgh while he was on a trip to America. Upon his arrival to the city, he was hurriedly shuttled off to the Edgar Thomson Works to view Carnegie's interpretation of a Spencerian paradise. After suffering exposure to the amenities of smoke, dirt, dust, heat, noise, odors, and such, which were all attendant to the Bessemer steelmaking process, Spencer's comment to Carnegie was: "Six months residence here would justify suicide."[52]

Table 8.5. U.S. Steel's Market Share of National Ingot Steel Production, 1901–1930

Year	% Market	Steel Shipped (Million tons)	Income (Million $)
1901	65.7	na	na
1902	65.2	8.9	90.3
1903	63.1	8.1	55.4
1904	60.6	7.3	30.2
1905	59.9	10.1	68.6
1906	57.7	11.3	98.1
1907	57.1	11.5	104.6
1908	55.9	6.8	45.7
1909	55.7	10.6	79.0
1910	54.3	11.8	87.4
1911	53.9	10.3	55.3
1912	54.1	13.8	54.2
1913	53.2	13.4	81.2
1914	50.3	9.9	23.4
1915	50.9	12.8	75.9
1916	48.9	17.1	271.5
1917	45.0	16.9	224.2
1918	44.0	15.6	125.3
1919	49.6	13.5	76.8
1920	45.8	15.5	109.7
1921	55.4	8.8	36.6
1922	45.2	13.1	39.6
1923	45.2	15.9	108.7
1924	43.4	12.7	85.1
1925	41.6	14.8	90.6
1926	42.0	15.8	116.7
1927	41.1	14.3	87.9
1928	39.0	15.4	114.1
1929	38.7	16.8	197.5
1930	41.2	12.8	104.4

Source: Kenneth Warren, *Big Steel: The First Century of the United States Steel Corporation* (Pittsburgh: University of Pittsburgh Press, 2001), created from data found in tables A3 and A9 on pages 359 and 363.

Carnegie was far from being a perfect man. He had little formal education. J. Edgar Thomson and Tom Scott were his superiors, as well as mentors at the Pennsylvania Railroad, and he kept a close association with them that continued for years after he resigned from their employ, finally ending with Thomson's death and the disintegration of Scott's friendship due to the Texas and Pacific Railroad affair. It is only reasonable, then, that his entrepreneurial philosophies would have evolved along lines that mimicked these men, to-

Chart 8.2. Note that U.S. Steel's market share (based on ingot production) followed a steadily declining curve beginning from the company's inception in 1901, with the exception of a brief span that included the WWI period. The company lost one-quarter of the market within the time frame of the data.

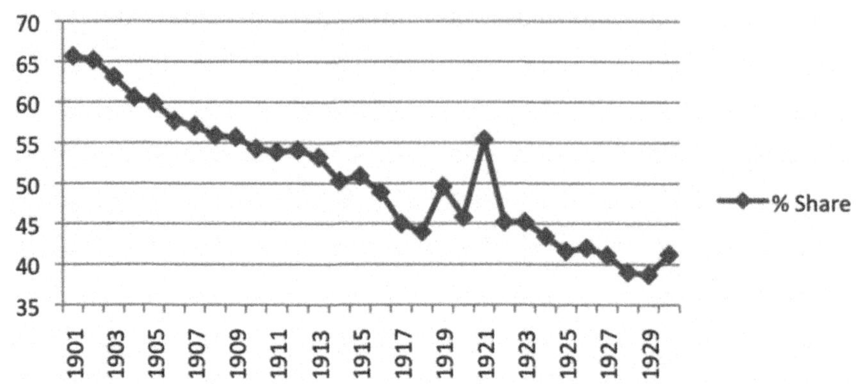

From table 8.5.

gether with others who were firmly ensconced at the very top of that institution at an especially critical period, during which the Pennsylvania Railroad was in the act of transforming into America's largest corporation. Carnegie was still an extremely impressionable young man at the time. He viewed his own father as a failure.[53] Many of the business policies he grew accustomed to and adopted would be considered illegal today. He had to learn to stand up to many men or be cut down in business. His personality was very conflicted through, among other things, his simultaneous beliefs in Spencer, a perceived role as the friend of the workingman (in his mind), and the need to defend his budding business empire.

Andrew Carnegie's early plan was to dedicate his annual income in excess of $50,000 to doing charitable works once it reached that point, although this scenario never quite came to fruition.[54] Yet he did transform an industry, and in that process he helped aid the rapid conversion of America into a world power. His business accomplishments thrust him into a position above all others. In the end, he was considered by many to be the richest man in the world, at least on a cash basis.[55] There are a number of instances in which he recognized the arrogance of his ways, such as his confrontation with Frick or his dealings with McLuckie (a leader of the 1892 Homestead strike), where he attempted in at least some small way to try to make amends for bad feelings from these past disputes. His unsuccessful peacemaking entreaty to Frick resulted in Frick's now famous "meet you in hell" response.

When it almost seemed as though there was nothing more that he wished to achieve, Carnegie gave away his vast fortune for the good of mankind.

Notes

INTRODUCTION

1. Kenneth Warren, *Wealth, Waste, and Alienation: Growth and Decline in the Connellsville Coke Industry* (Pittsburgh: University of Pittsburgh Press, 2001), 29.
2. Victor S. Clark, *History of Manufacturers in the United States, Volume II, 1860–1893* (New York: McGraw-Hill, 1929), 237.

CHAPTER 1. WROUGHT IRON AND PUDDLING

1. O. F. Hudson and Guy D. Bengough, *Iron and Steel: An Introductory Textbook for Engineers and Metallurgists* (London: Constable and Company Limited, 1913), 49.
2. H. O. Hofman and R. H. Richards, *An Outline of the Metallurgy of Iron and Steel* (Boston: Massachusetts Institute of Technology, 1904), 126.
3. Hofman and Richards, *An Outline of the Metallurgy of Iron and Steel.*
4. Thomas Turner, *The Metallurgy of Iron and Steel* (London: Charles Griffin and Company Limited, 1895), 15.
5. Hofman and Richards, *An Outline of the Metallurgy of Iron and Steel.*
6. James Aston and Edward B. Story, *Wrought Iron: Its Manufacture, Characteristics and Applications*, second edition (Pittsburgh: A. M. Byers Co., 1947), 11–12.
7. James M. Swank, *Introduction to a History of Ironmaking and Coal Mining in Pennsylvania* (Philadelphia: published by the author, 1878), 55.
8. I extrapolated the location of Plumsock by first identifying the location of Middleton (Middletown by Swank in reference 9) on an 1826 map of Pennsylvania. I then used the town of New Salem and the National Road or National Pike to Brownsville on that map to identify the meanders in Redstone Creek that would be consistent with Plumsock (Middleton). This evidence was transferred from the 1826 map to be used as guidelines on a modern-day Google Map of the region, on which neither Plumsock

nor Middleton exists, but both the town of New Salem and the National Pike (present-day U.S. Route 40) do. Doing so indicates that Plumsock was approximately located slightly southeast (about one-half mile) of present-day Waltersburg on Redstone Creek. A modern crossroads located at this area of Redstone Creek is known as PA State Route 51 (Pittsburgh Road) at State Route 4010, which is also known as Upper Middletown Road, among other names.

9. Robert B. Gordon, *American Iron, 1607–1900* (Baltimore and London: Johns Hopkins University Press, 1996), 134.

10. John L. Bray, *Ferrous Production Metallurgy* (New York: John Wiley & Sons, 1942), 202.

11. James J. Davis, *The Iron Puddler: My Life in the Rolling Mills and What Came of It* (New York: Grosset & Dunlap Publishers, New York), 1922, 90.

12. Frederick Overman, *The Manufacture of Iron in All Its Various Branches*, second edition (Philadelphia: Henry C. Baird, 1851), 266.

13. Orin W. McMullan and Albert M. Talbot, *Metallurgy of Iron* (Scranton: International Textbook Co., 1936), 55–56; Bray, *Ferrous Production Metallurgy*; Bradley Stoughton, *The Metallurgy of Iron and Steel*, second edition (New York: McGraw-Hill Book Co., 1911), 59.

14. Davis, *The Iron Puddler*, 101.

15. Davis, *The Iron Puddler*, 97.

16. Davis, *The Iron Puddler*, 99.

17. Bray, *Ferrous Production Metallurgy*.

18. Ernest J. Teichert, *Ferrous Metallurgy, volume 1* (Pennsylvania: The Pennsylvania State College, 1941), 421–27; Gordon, *American Iron, 1607–1900*, 142.

19. Gordon, *American Iron, 1607–1900*.

20. Bray, *Ferrous Production Metallurgy*, 203.

21. Davis, *The Iron Puddler*, 105.

22. Teichert, *Ferrous Metallurgy*, 424.

23. Bray, *Ferrous Production Metallurgy*; Teichert, *Ferrous Metallurgy*.

24. Davis, *The Iron Puddler*, 106–7.

25. Davis, *The Iron Puddler*, 110.

26. Hugh P. Tiemann, *Iron and Steel: A Pocket Encyclopedia*, second edition (New York: McGraw-Hill Book Co., Inc., 1919), 376; Teichert, *Ferrous Metallurgy*, 425.

27. Aston and Story, *Wrought Iron: Its Manufacture, Characteristics and Applications*, 8.

28. Teichert, *Ferrous Metallurgy*, 423.

29. Aston and Story, *Wrought Iron: Its Manufacture, Characteristics and Applications*, 34.

CHAPTER 2. CRUCIBLE STEEL

1. Harry Brearley, *Steel-Makers and Knotted String* (London: Longmans, Green and Co., 1933), 5–6.

2. John Percy, *Metallurgy: Iron and Steel* (London: John Murray, 1864), 774.

3. Frederico Giolitti, *The Cementation of Iron and Steel*, translated from the Italian by Joseph Richards and Charles Rouiller (New York: McGraw-Hill Book Co., Inc., 1915), 210–12.

4. Kenneth C. Barraclough, *Steelmaking Before Bessemer, Vol. 2, Crucible Steel: The Growth of Technology* (London: The Metals Society, 1984), 44–45.

5. Percy, *Metallurgy*, 795.

6. Percy, *Metallurgy*, 859.

7. Giolitti, *The Cementation of Iron and Steel*, 222.

8. Brearley, *Steel-Makers and Knotted String*, 6–7.

9. Barraclough, *Steelmaking Before Bessemer, Vol. 2*, 43.

10. Barraclough, *Steelmaking Before Bessemer, Vol. 2*, 8.

11. Brearley, *Steel-Makers and Knotted String*, 5.

12. Fred Osborn, *The Story of the Mushets* (London: Thomas Nelson and Sons Ltd., 1952), 28–29.

13. Barraclough, *Steelmaking Before Bessemer, Vol. 2*, 33–39.

14. J. M. Camp and C. B. Francis, *Making, Shaping and Treating of Steel*, 4th edition, (Pittsburgh: Carnegie Steel Co., 1925), 247.

15. James M. Swank, *The Manufacture of Iron in All Ages* (Philadelphia: Published by the Author, 1884), 296.

16. Swank, *The Manufacture of Iron in All Ages*, 297; Barraclough, *Steelmaking Before Bessemer, Vol. 2*, 219.

17. Barraclough, *Steelmaking Before Bessemer, Vol. 2*, 220.

18. Victor S. Clark, *History of Manufactures in the United States, Volume 1, 1607–1860* (New York: McGraw Hill Book Co., 1929), 517.

19. Swank, *The Manufacture of Iron in All Ages*, 297–98.

20. Clark, *History of Manufactures in the United States*, 517.

21. Barraclough, *Steelmaking Before Bessemer, Vol. 2*, 220.

22. Henry Marion Howe, *The Metallurgy of Steel* (New York and London: The Scientific Publishing Co., 1896), 304.

23. Camp and Francis, *Making, Shaping and Treating of Steel*, 250–51.

24. Camp and Francis, *Making, Shaping and Treating of Steel*, 253.

25. Camp and Francis, *Making, Shaping and Treating of Steel*, 250.

26. Howe, *The Metallurgy of Steel*, 304.

27. Howe, *The Metallurgy of Steel*, 304–5.

28. Brearley, *Steel-Makers and Knotted String*, 66–68.

29. Howe, *The Metallurgy of Steel*, 305.

30. John L. Bray, *Ferrous Production Metallurgy*, 2nd Printing (New York: John Wiley & Sons, 1947), 218.

31. Camp and Francis, *Making, Shaping and Treating of Steel*, 250.

32. Brearley, *Steel-Makers and Knotted String*, 69–75.

CHAPTER 3. FUELS AND TRANSPORTATION

1. Howard N. Eavenson, *The First Century and a Quarter of American Coal Industry* (Pittsburgh: Privately Printed, 1942), 22.

2. Eavenson, *The First Century and a Quarter of American Coal Industry*, 155.

3. Eavenson, *The First Century and a Quarter of American Coal Industry*, 162–63.

4. Eavenson, *The First Century and a Quarter of American Coal Industry*, 171–72.

5. Eavenson, *The First Century and a Quarter of American Coal Industry*, 166.

6. Eavenson, *The First Century and a Quarter of American Coal Industry*, 187–88.

7. Pennsylvania Historical and Museum Commission, Historical Marker located at Grandview Avenue between Ulysses and Bertha Streets, Pittsburgh.

8. Charles Henry Ambler, *A History of Transportation in the Ohio Valley* (Glendale, CA: The Arthur H. Clark Co., 1932), Pittsburgh Digital Research Library online edition, 310.

9. Walter R. Wagner, Louis Heyman, Richard E. Grey, David J. Belz, Richard Lund, Addison S. Cate, and Curtis Edgerton, *Geology of the Pittsburgh Area, Pennsylvania Geological Survey*, Harrisburg, Pennsylvania, 1970, 18.

10. Reports from Commissioners, "Inspectors and Others, Education Department" (Philadelphia International Exhibition) Vol. XXXVI, Session 8 (February–August 14, 1877), 304–5.

11. Pennsylvania, *Annual Report of the Geological Survey of Pennsylvania for 1886*, Board of Commissioners for the Geological Survey, Harrisburg, Pennslyvania, 1887, 27–28.

12. Joseph D. Weeks, *Report on the Manufacture of Coke, U.S. Census for 1880* (New York: David Williams, 1885), 19.

13. Weeks, *Report on the Manufacture of Coke*, 23.

14. Frederick Overman, *The Manufacture of Iron in All Its Various Branches* (Philadelphia: Henry C. Baird, 1851), 119–26.

15. Overman, *The Manufacture of Iron in All Its Various Branches*.

16. Weeks, *Report on the Manufacture of Coke*, 86.

17. Weeks, *Report on the Manufacture of Coke*, 89.

18. Weeks, *Report on the Manufacture of Coke*.

19. Edward V. D'Invilliers, "Estimated Cost of Mining and Coking and Relative Commercial Returns from Operating in the Connellsville and Watson-Reynoldsville Districts, Pennsylvania," *Transactions of the American Institute of Mining Engineers* XXXV, 1905, 48.

20. F. C. Keighley, The Connellsville Coke Region, *The Engineering Magazine* XX (October 1900 to March 1901): 26.

21. James M. Swank, *Introduction to a History of Ironmaking in Pennsylvania Contributed to the Final Report of the Pennsylvania Board of Centennial Managers* (Philadelphia: Published by the Author, 1878), 114.

22. John Fulton, *Coke: A Treatise in the Manufacture of Coke and the Savings of By-Products* (Scranton: The Colliery Engineer Co., 1895), 39.

23. Weeks, *Report on the Manufacture of Coke*, 7.

24. Overman, *The Manufacture of Iron in All Its Various Branches*, 125.

25. *Annual Report of the Geological Survey of Pennsylvania for 1886*, 27.

26. Weeks, *Report on the Manufacture of Coke*, 26.

27. Eavenson, *The First Century and a Quarter of American Coal Industry*, 581.

28. American Society of Mechanical Engineers, Souvenir of the Spring Meeting of the ASME, Pittsburgh, March 1928, 111.

29. H. H. Lowry, *Chemistry of Coal Utilization*, volume II (New York: John Wiley & Sons, 1945), 1653.

30. Lowry, *Chemistry of Coal Utilization*, 1586–87.

31. Murraysville Gas Well Historical Marker, Explore PA History.com, http://www.explorepahistory.com/hmarker.php?markerId=436.

32. *Annual Report of the Geological Survey of Pennsylvania for 1886*, 19.

33. George H. Thurston, *Pittsburgh's Progress, Industries and Resources* (Pittsburgh: A. A. Anderson & Son, 1886), 7.

34. William Bender Wilson, *History of the Pennsylvania Railroad Company with Plan of Organization, Portraits of Officials and Biographical Sketches in Two Volumes*, volume 1 (Philadelphia: Henry T. Coats & Company, 1899), 216.

35. Photocopy of an article found in Pittsburgh Rivers file folder, Carnegie Library of Pittsburgh, Pennsylvania Department.

36. Historical Timeline 1818–1820, Construction of the National Road 1806–1834, http://www.nationalroadpa.org (timeline).

37. James Veech, *A History of the Monongahela Navigation Company by an Original Stockholder* (Pittsburgh: Bakewell and Marthens, 1873), 7.

38. *Annual Report of the Geological Survey of Pennsylvania for 1886*, 22.

39. *Annual Report of the Geological Survey of Pennsylvania for 1886*, 23.

40. *Pittsburgh Post-Gazette*, "A Fact a Day About Pittsburgh," October 3, 1922, file folder Pittsburgh Rivers, Carnegie Library of Pittsburgh, Pennsylvania Department.

41. From a postcard notation of a painting of Port Perry by Howard Fogg, distributed by the Pittsburgh & Lake Erie Railroad Company, c. 1960s.

CHAPTER 4. IRON

1. Robert H. Thurston, *A Text-Book of the Materials of Construction, For Use in Engineering and Technical Schools* (New York: John Wiley & Sons, 1885), 44–49.

2. Thomas Turner, *The Metallurgy of Iron* (London: Charles Griffin & Co. Ltd., 1908), 8–10.

3. John A. Ricketts, *History of Ironmaking, Iron & Steel Society* (Warrendale, PA, 2000), 24.

4. Frederick Overman, *A Treatise on Metallurgy*, third edition (New York: D. Appleton & Co., 1855), 360.

5. Overman, *A Treatise on Metallurgy*, 522.

6. Malcolm Keir, *Manufacturing Industries in America* (New York: The Ronald Press Company, 1920), 121.

7. John Percy, *Metallurgy: Iron and Steel* (London: John Murray, 1864), 257–61.

8. Turner, *The Metallurgy of Iron*, 10–11; Percy, *Metallurgy: Iron and Steel*, 888–89.

9. William F. Durfee, "The Development of American Industries Since Columbus," *Popular Science Monthly* XXXVIII (November 1890 to April 1891), D. Appleton and Company, New York, 459–60.

10. Frederick Overman, *The Manufacture of Iron in All Its Various Branches*, second Edition (Philadelphia: Henry C. Baird, 1851), 177–78.

11. Bradley Stoughton, *The Metallurgy of Iron and Steel*, second edition (New York: McGraw-Hill Book Co., 1911), 40.

12. Percy, *Metallurgy: Iron and Steel*, 394–95.

13. Percy, *Metallurgy: Iron and Steel*, 397.

14. Turner, *The Metallurgy of Iron*, 20.

15. Percy, *Metallurgy: Iron and Steel*, 421–22.

16. Overman, *The Manufacture of Iron in All Its Various Branches*, 530–31.

17. Ricketts, *History of Ironmaking, Iron & Steel Society*, 58.

18. J. E. Johnson Jr., *Blast-Furnace Construction in America* (New York: McGraw-Hill Book Co., 1917), 226.

19. James M. Swank, *Introduction to a History of Ironmaking and Coal Mining in Pennsylvania* (Philadelphia: Published by the Author, 1878), 73–76; *Annual Report of the Secretary of Internal Affairs of the Commonwealth of Pennsylvania, Part III Industrial Statistics*, Vol. X, 1881–1882, Lane S. Hart State Printer & Binder, 1883, 30–32.

20. Overman, *A Treatise on Metallurgy*, 519.

21. Swank, *Introduction to a History of Ironmaking*, 58–59. *Report of the Secretary of Internal Affairs of the Commonwealth of Pennsylvania*, 26–28.

22. Swank, *Introduction to a History of Ironmaking*, 60.

23. Keir, *Manufacturing Industries in America*, 115.

24. Swank, *Introduction to a History of Ironmaking*, 60.

25. J. P. Lesley, *The Iron Manufacturer's Guide to the Furnaces, Forges and Rolling Mills of the United States* (New York: John Wiley, 1859), 248–49.

26. Swank, *Introduction to a History of Ironmaking*, 61.

27. Durfee, "The Development of American Industries Since Columbus," 462–63.

28. Durfee, "The Development of American Industries Since Columbus," 461.

29. Joseph D. Weeks, *Report on the Manufacture of Coke* (New York: David Williams, 1885), 27.

30. Swank, *Introduction to a History of Ironmaking*, 71. Weeks, *Report on the Manufacture of Coke*, 23.

31. John Fulton, *Coke: A Treatise on the Manufacture of Coke and the Saving of By-Products* (Scranton: The Colliery Engineer Co., 1895), 94.

32. F. C. Kieghley, *History of Connellsville Coke, The Foundry* XVI, no. 96 (August 1990), The Foundry Publishing Company, Detroit, 38–40.

33. American Iron and Steel Association, *Directory of the Iron and Steel Works of the United States*, Philadelphia, 1880, 34–35. Commonwealth of Pennsylvania, Second Annual Report of the Bureau of Statistics of Pennsylvania for the Years 1873–74, B. F. Meyers State Printer, 1875, 262–63.

34. Swank, *Introduction to a History of Ironmaking*, 78.

35. J. E. Johnson Jr., *Blast-Furnace Construction in America* (New York: McGraw-Hill Book Company, Inc., 1917), 6.

36. James M. Swank, *History of the Manufacture of Iron in All Ages*, second edition (Philadelphia: The American Iron and Steel Association, 1892), 454.

37. Charles Reitell, *Machinery and Its Benefits to Labor in the Crude Iron and Steel Industries*, Doctoral Thesis, University of Pennsylvania, The Collegiate Press, George Banta Publishing Company, Menasha, Wisconsin, 1917, 16.

38. Reitell, *Machinery and Its Benefits to Labor*.

39. Archibald P. Head, *Notes on American Iron and Steel Practice, Excerpt of Proceedings of the Meeting of the South Staffordshire Institute of Iron & Steel Works' Managers*, Dudley, Brierley-Hill, March 19, 1898, 8.

40. LaVerne W. Spring, *Non-Technical Chats on Iron and Steel* (New York: Frederick A. Stokes Co., 1917), 61.

41. Reitell, *Machinery and Its Benefits to Labor*, 10–21.

42. American Iron and Steel Association, *Directory of the Iron and Steel Works of the United States*.

43. American Iron and Steel Association, *Directory of the Iron and Steel Works of the United States*, Philadelphia.

44. James Howard Bridge, *The Inside History of the Carnegie Steel Company* (New York: The Aldine Book Company, 1903), 54–55.

45. Bridge, *The Inside History of the Carnegie Steel Company*.

46. Andrew Carnegie, *Autobiography of Andrew Carnegie* (Boston and New York: Houghton Mifflin Company, 1920), 131–32.

47. James Park and James Hemphill, "Kloman Eulogy," *Proceedings of the Engineers' Society of Western Pennsylvania* 1 (January 1880 to June 1881), Pittsburgh, 1882, 253.

48. Herbert N. Casson, *The Romance of Steel* (New York: A. S. Barnes & Co., 1907), 80.

49. Joseph Frazier Wall, *Andrew Carnegie* (New York: Oxford University Press, 1970), 323–24.

50. The Lucy Furnace, Pittsburgh, *The Iron Age* XII, no. 6, 5.

51. John B. Pearse, *A Concise History of the Iron Manufacture of the American Colonies* (Philadelphia: Allen, Lane & Scott, 1876), 140.

52. *Report by Her Majesty's Secretaries of Embassy and Legation on the Manufactures, Commerce, &c., of the Countries in Which They Reside* (Commercial No. 16-1874), Part I, Harrison and Sons, London, 1874, 247–48.

53. William B. Sipes, *The Pennsylvania Railroad: Its Origin, Construction, Condition and Connections, The Passenger Department (PRR)*, Philadelphia, 1875, 216.

54. A. L. Holley and Lenox Smith, *American Iron and Steel Works* XXXII, The Lucy Furnaces, Messrs. Carnegie Brothers and Co., Pittsburgh, PA, Engineering, November 28, 1878, 407.

55. Bridge, *The Inside History of the Carnegie Steel Company*, 65.

56. Carnegie, *Autobiography of Andrew Carnegie*, 182. W. Paul Strassmann, *Risk and Technological Innovation: American Manufacturing Methods during the Nineteenth Century* (New York: Cornell University Press, 1959), 52–53.

57. Bridge, *The Inside History of the Carnegie Steel Company*, 64.

58. Bridge, *The Inside History of the Carnegie Steel Company*, 62–63.

59. Wall, *Andrew Carnegie*, 323–24.

60. Casson, *The Romance of Steel*, 88.

61. "Report on the Progress of the Iron and Steel Industries in Foreign Countries," *Journal of the Iron & Steel Institute*, London, 1875, 655.

62. Bridge, *The Inside History of the Carnegie Steel Company*, 56.

63. Wall, *Andrew Carnegie*.

64. Holley and Smith, *American Iron and Steel Works*.

65. I. Lowthian Bell, *Mr. I. Lowthian Bell and the Blair Direct Process* (Pittsburgh: James M'Millin, 1875), 20.

66. James Gayley, "The Development of American Blast-Furnaces, With Special Reference to Large Yields," *Transactions of the American Institute of Mining Engineers* XIX, New York, 1891, 932–34.

67. "American vs. British Pig Iron Making," *The Foundry* XVI, no. 96 (August 1900), The Foundry Publishing Co., Detroit, 253.

68. Joseph Nimmo Jr., "Report of the Internal Commerce of the United States for the Fiscal Year 1881–1882," Government Printing Office, Washington, DC, 1884, 155.

CHAPTER 5. THE BESSEMERS ARRIVE AT BRADDOCK

1. Stewart H. Holbrook, *Iron Brew: A Century of American Iron and Steel* (New York: Macmillan Company, 1939), 2–4.

2. LaVerne W. Spring, *Non-Technical Chats on Iron and Steel* (New York: Frederick A. Stokes Co., 1917), 124.

3. Henry Bessemer, *Sir Henry Bessemer, F. R. S.: An Autobiography* (London: Offices of Engineering, 1905), 135–36.

4. Bessemer, *Sir Henry Bessemer*, 138–50.

5. Bessemer, *Sir Henry Bessemer*, 154–56.

6. Carnegie Illinois Steel Corporation, "Metallurgy of Iron and Steel: Graduate Work in Industry," Carnegie Illinois Steel Corporation, Pittsburgh, 1945, F. L. Toy, Section 2.4.1, 30.

7. Robert Hunt, *Ure's Dictionary of Arts, Manufactures, and Mines*, seventh edition, Vol. III (London: Longmans, Green, and Co., 1878), 903.

8. Thomas Turner, *The Metallurgy of Iron and Steel*, vol. 1 (London: Charles Griffin & Co. Ltd., 1895), 35–39; David Kirkaldy, *Results of an Experimental Inquiry into the Tensile Strength and Other Properties of Various Kinds of Wrought-Iron and Steel* (London: Published by the Author, 1866), 36.

9. John Percy, *Metallurgy: Iron and Steel* (London: John Murray, 1864), 814.

10. Carnegie Illinois Steel Corporation, "Metallurgy of Iron and Steel," 31.

11. Percy, *Metallurgy: Iron and Steel*.

12. Alexander L. Holley, *The Bessemer Process and Works in the United States*, Reprinted from the *Troy Dailey Times* of July 27, 1868 (New York: D. Van Nostrand, 1868), 10–11.

13. Percy, *Metallurgy: Iron and Steel*.

14. Jeanne McHugh, *Alexander Holley and the Makers of Steel* (Baltimore: Johns Hopkins University Press, 1980), 158–62; Turner, *The Metallurgy of Iron and Steel*,

40; James M. Swank, *History of the Manufacture of Iron in All Ages* (Philadelphia: Published by the Author, 1884), 314.

15. Alexander L. Holley, *Bessemer Machinery: A Lecture, Delivered Before the Students of the Stevens Institute of Technology* (Philadelphia: Merrihew & Sons, 1873), 7.

16. John N. Boucher, *William Kelly: A True History of the So-Called Bessemer Process* (Greensburg, PA: Published by the Author, 1924), 81.

17. Robert W. Hunt, "A History of the Bessemer Manufacture in America," *Transactions of the American Institute of Mining Engineers* 5, May 1876 to February 1877, 201.

18. Percy, *Metallurgy: Iron and Steel*, 824.

19. Percy, *Metallurgy: Iron and Steel*, 823.

20. Percy, *Metallurgy: Iron and Steel*.

21. Holley, *Bessemer Machinery*, 10–11.

22. Holley, *Bessemer Machinery*, 12.

23. W. F. Durfee, "An Account of the Experimental Steel Works at Wyandotte, Michigan," *Transactions of the American Society of Mechanical Engineers* 6, November 1884 and May 1885, New York, 40–45; James M. Swank, *Introduction to a History of Ironmaking and Coal Mining in Pennsylvania* (Philadelphia: Published by the Author, 1878), 84.

24. Swank, *History of the Manufacture of Iron in All Ages*, 308–9.

25. Durfee, "An Account of the Experimental Steel Works at Wyandotte, Michigan."

26. Robert W. Hunt, "The Original Bessemer Steel Plant at Troy," *Transactions of the American Society of Mechanical Engineers* 6, November 1884 and May 1885, New York, 61–65.

27. John B. Pearse, *A Concise History of the Iron Manufacture of the American Colonies* (Philadelphia: Allen, Lane & Scott, 1876), 195; Robert W. Hunt, "A History of Bessemer Manufacture in America," 206; *The Bulletin* 16, December 6, 1882, American Iron and Steel Association, 325.

28. *Reports by Her Majesty's Secretaries of Embassy and Legation of the Manufactures, Commerce, &c., of the Countries in Which They Reside, Part I* (Commercial No. 16 [1874]), Presented to Both Houses of Parliament by Command of Her Majesty, June 1874, London, Printed by Harrison and Sons, 1874, 370; Robert W. Hunt, "A History of Bessemer Manufacture in America," 215.

29. Henry M. Howe, "Notes on the Bessemer Process," *Transactions of the American Institute of Mining Engineers* 19, New York, 1891, 1150.

30. *Reports by Her Majesty's Secretaries*, 370–2.

31. Swank, *Introduction to a History of Ironmaking and Coal Mining in Pennsylvania*, Ibid, 83–84.

32. John F. Winslow, John A. Griswold, and Daniel J. Morrell, *The Pneumatic or Bessemer Process of Making Iron and Steel* (Philadelpha: Self-Published, n.d.), but c. 1869, 25–26. This was a business brochure for promoting the sale of the product.

33. Albert Sauveur, ed., *The Iron and Steel Magazine* XI, January to June 1906, Cambridge, Massachusetts, 135; from an article in *The Bulletin* of the American Iron and Steel Association, January 20, 1906.

34. W. Paul Strassmann, *Risk and Technological Innovation* (Ithaca: Cornell University Press, 1959), 34.

35. New American Supplement to the *Encyclopedia Britannica*, Vol. IV (New York: The Werner Company, 1897), 2493; Swank, *History of the Manufacture of Iron in All Ages*, 85; Strassman, *Risk and Technological Innovation*.

36. Turner, *The Metallurgy of Iron and Steel*, 41.

37. J. S. Jeans, *Steel: Its History, Manufacture, Properties and Uses* (London: E. & F. N. Spon, London, 1880), 145–46; *Encyclopedia Britannica*, Vol. IV.

38. *Twentieth Annual Report of the Pennsylvania Railroad Co.*, Philadelphia, 1867, 63–64.

39. Swank, *History of the Manufacture of Iron in All Ages*, 84–85; Pearse, *A Concise History of the Iron Manufacture of the American Colonies*, 195–96.

40. Pearse, *A Concise History of the Iron Manufacture of the American Colonies*, 197.

41. *Reports by Her Majesty's Secretaries*, 378.

42. Memorial of Alexander Lyman Holley, C.E., LL.D., The American Institute of Mining Engineers, New York, 1884, 69.

43. *Reports by Her Majesty's Secretaries*, 377–78.

44. Memorial of Alexander Lyman Holley, 23–25.

45. *Reports by Her Majesty's Secretaries*, 375–79.

46. Winslow, Griswold, and Morrell, *The Pneumatic or Bessemer Process of Making Iron and Steel*, 9.

47. *Reports by Her Majesty's Secretaries*, 381–82.

48. *Reports by Her Majesty's Secretaries*, 387.

49. J. Edgar Thomson to Andrew Carnegie, Letter, Nov. 14, 1872, Library and Archives Division, Historical Society of Western Pennsylvania, Pittsburgh, Pennsylvania, Records of the Carnegie Steel Company, MSS#315, Box 36, FF.

50. Albert J. Churella, *The Pennsylvania Railroad*, volume 1 (Philadelphia: University of Pennsylvania Press, 2013), 212.

51. Joseph Frazier Wall, *Andrew Carnegie* (New York: Oxford University Press, 1970), 113–21.

52. Wall, *Andrew Carnegie*, 302.

53. Wall, *Andrew Carnegie*, 302–6.

54. James Howard Bridge, *The Inside History of the Carnegie Steel Company* (New York: The Aldine Book Company, 1903), 76–77.

55. Wall, *Andrew Carnegie*.

56. Herbert N. Casson, *The Romance of Steel: The Story of a Thousand Millionaires* (New York: A. S. Barnes & Company, 1907), 21–23; Bridge, *The Inside History of the Carnegie Steel Company*, 78–79; Obituary William Richard Jones, *The Journal of the Iron and Steel Institute*, E. F. & N. Spon, London 1890, 179–80; R. W. Raymond, "Biographical Notice of William R. Jones," *Transactions of the American Institute of Mining Engineers* XVIII, May 1889–February 1890, 621–22.

57. Bridge, *The Inside History of the Carnegie Steel Company*, 78–79.

58. *The Journal of the Iron and Steel Institute*.

59. Wall, *Andrew Carnegie*, 100.

60. *The American Monthly Review of Reviews* XXIII, January–June 1901, New York, 466.

61. Wall, *Andrew Carnegie*, 163–68; *The American Monthly Review of Reviews*.

62. A. L. Holley to Andrew Carnegie, Letter, May 4, 1874, Library and Archives Division, Historical Society of Western Pennsylvania, Pittsburgh, Pennsylvania, Records of the Carnegie Steel Company, MSS#315, Box 36A, FF 1.

63. Wall, *Andrew Carnegie*, 343.

64. "Transactions," *American Society of Mechanical Engineers* XIII, New York, 1892, 50; Robert W. Hunt, "A History of the Bessemer Manufacture in America," 206.

65. A. L. Holley to Andrew Carnegie.

66. Wall, *Andrew Carnegie*, 320–21.

67. Casson, *The Romance of Steel*, 25.

68. William P. Shinn, "Pittsburgh: Its Resources and Surroundings," *Transactions of the American Institute of Mining Engineers* VIII, May 1879 to February 1880, Easton, Pennsylvania, 1880, 18.

69. *Directory to the Iron and Steel Works of the United States* (Philadelphia: The American Iron and Steel Association, 1880), 116; *Directory to the Iron and Steel Works of the United States* (Philadelphia: The American Iron and Steel Association, 1882), 161.

70. Henry Marion Howe, *The Metallurgy of Steel* (Philadelphia: The Scientific Publishing Company), 1904 edition, 320.

71. Wall, *Andrew Carnegie*, 315.

72. Wall, *Andrew Carnegie*, 345.

73. Bridge, *The Inside History of the Carnegie Steel Company*, 81–82; Casson, *The Romance of Steel*, 27–28.

74. Bridge, *The Inside History of the Carnegie Steel Company*, 79–80; Casson, *The Romance of Steel*, 31–33; F. C. Johnson, ed., *The Historical Record* III, no. 3 (July 1889), Press of the Historical Record, Wilkes-Barre, Pennsylvania, 1889, 18–19.

75. Wall, *Andrew Carnegie*.

76. *Transactions of the American Society of Mechanical Engineers* X, American Society of Mechanical Engineers, New York, 1889, 842.

77. Wall, *Andrew Carnegie*, 527.

78. Wall, *Andrew Carnegie*, 520–21; W. R. Jones, "The Manufacture of Bessemer Steel and Steel Rails in the United States," *The Engineering and Mining Journal* XXI, June 4, 1881, Scientific Publishing Company, New York, 382–83.

79. Jones, "The Manufacture of Bessemer Steel and Steel Rails in the United States," 383.

80. Bridge, *The Inside History of the Carnegie Steel Company*, 172.

81. Wall, *Andrew Carnegia*, 527.

82. Wall, *Andrew Carnegie*, 521–22.

83. Casson, *The Romance of Steel*; Bridge, *The Inside History of the Carnegie Steel Company*, 90–91.

84. William Powell Shinn, Obituary, *Proceedings of the American Society of Civil Engineers* XVIII, New York, 1892, 124; *Minutes of the Proceedings of the Institution of Civil Engineers* CXI, London, 1893, 391–92.

85. Bridge, *The Inside History of the Carnegie Steel Company*, 84–85.

86. Strassmann, *Risk and Technological Innovation*, 23.

87. *Minutes of the Proceedings of the Institution of Civil Engineers*, 393; William P. Shinn, "On Railroad Accounts and Returns," *Transactions of the American Society of Civil Engineers* V, New York, 1876, 215–18.

88. Wall, *Andrew Carnegie*, 525–26; Bridge, *The Inside History of the Carnegie Steel Company*, 94–95.

89. Bridge, *The Inside History of the Carnegie Steel Company*, 84–86.

90. Bridge, *The Inside History of the Carnegie Steel Company*, 113.

91. Wall, *Andrew Carnegie*, 344.

92. Wall, *Andrew Carnegie*, 321.

93. Bridge, *The Inside History of the Carnegie Steel Company*, 104–5.

94. "Edgar Thomson Works: Mother Steel Plant of Carnegie Steel," *US Steel News* vol. I, no. 4 (September 1936): 16–18.

95. Wall, *Andrew Carnegie*, 359.

96. Howe, *The Metallurgy of Steel*, 317 and 323.

97. Howe, *The Metallurgy of Steel*, 323.

98. Howe, *The Metallurgy of Steel*.

99. Howe, *The Metallurgy of Steel*.

100. Howe, *The Metallurgy of Steel*, 321; "Use of Steam in the Bessemer Process," *The Journal of the Iron and Steel Institute*, E. & F. N. Spon, London, 1883, 747.

101. Howe, "Notes on the Bessemer Process," 1156.

102. Bridge, *The Inside History of the Carnegie Steel Company*, 93.

103. "The Edgar Thomson Steel Works," *The Engineering and Mining Journal* XXI, Scientific Publishing Company, New York, 1876, 199.

104. Captain W. R. Jones, "On the Manufacture of Bessemer Steel and Steel Rails in the United States," *The Journal of the Iron and Steel Institute*, E. & F. N. Spon, London 1881, 133.

105. Captain W. R. Jones, "On the Manufacture of Bessemer Steel and Steel Rails in the United States," 134.

106. Henry M. Howe, "Notes on the Bessemer Process," 1128.

107. Captain W. R. Jones, "On the Manufacture of Bessemer Steel and Steel Rails in the United States," 130 and 134.

108. K. Mathieson, *How We Saw the United States of America* (Privately Printed, 1883), 23.

109. Captain W. R. Jones, "On the Manufacture of Bessemer Steel and Steel Rails in the United States."

110. Henry M. Howe, "Notes on the Bessemer Process," 1120.

111. *Journal of the Iron and Steel Institute* 17, 713–14.

112. Private communication with Kenneth Warren.

113. Jeans, *Steel: Its History, Manufacture, Properties and Uses*, 146–47.

114. Wall, *Andrew Carnegie*, 329–31.

115. Andrew Carnegie, *Autobiography of Andrew Carnegie* (New York: Houghton Mifflin Company, 1920), 101.

116. Wall, *Andrew Carnegie*, 331–33.

117. Raymond, "Biographical Notice of William R. Jones," 622; Obituary William Richard Jones, 180.

118. Park Benjamin, ed., *Modern Mechanism, Supplement to Appletons' Cyclopedia of Applied Mechanics* (New York: D. Appleton and Company, 1892), 807.

119. L. G. Laureau, *A Bessemer Converting House without a Casting-Pit, Reports on the Philadelphia International Exhibition of 1876, Vol. 1*, Reported to Both Houses of Parliament by Her Majesty, Her Majesty Stationary Office, London, 1877, 700–6.

120. Bridge, *The Inside History of the Carnegie Steel Company*, 87–88.

121. James Gayley, "The Development of American Blast Furnaces with Special Reference to Large Yields," *Transactions of the American Institute of Mining Engineers* XIX, New York, 1891, 936; Bridge, *The Inside History of the Carnegie Steel Company*, 87–88; "Greater Pittsburgh and Allegheny County Illustrated," *The American Manufacturer and Iron World*, 1901, for distribution at the Pan American Exposition, Buffalo, New York, 1901; biography of Julian Kennedy, no page number.

122. Julian Kennedy, "Blast Furnace Working," *Transactions of the American Institute of Mining Engineers* VIII, New York, 1880, 348.

123. William P. Shinn, "The Genesis of the Edgar Thomson Blast-Furnaces," *Transactions of the American Institute of Mining Engineers* XIX, New York, 1891, 675–76.

124. Julian Kennedy, "Hot-Blast Stoves at the Edgar Thomson Furnaces, 'D' and 'E,'" *Transactions of the American Institute of Mining Engineers* X, New York, 1882, 495–97.

125. Kennedy, "Blast-Furnace Working," 353–54.

126. Gayley, "The Development of American Blast Furnaces with Special Reference to Large Yields," 937; Kennedy, "Hot-Blast Stoves at the Edgar Thomson Furnaces," 351.

127. Shinn, "The Genesis of the Edgar Thomson Blast-Furnaces."

128. Shinn, "The Genesis of the Edgar Thomson Blast-Furnaces."

129. Gayley, "The Development of American Blast Furnaces with Special Reference to Large Yields," 939–40.

130. Gayley, "The Development of American Blast Furnaces with Special Reference to Large Yields," 932.

131. J. Stephen Jeans, ed., *Reports of the Commissioners—British Iron Trade Association—Iron, Steel, and Allied Industries of the United States*, Offices of the British Iron Trade Association, London, 1902, 400; Victor S. Clark, *History of Manufactures in the United States, Volume II 1860–1893* (New York: Carnegie Institution of Washington, McGraw Hill Book Company, Inc., 1929), 254.

132. Gayley, "The Development of American Blast Furnaces with Special Reference to Large Yields," 939–55.

133. Gayley, "The Development of American Blast Furnaces with Special Reference to Large Yields," 977.

134. Gayley, "The Development of American Blast Furnaces with Special Reference to Large Yields," 955.

135. E. C. Potter, "Review of American Blast Furnace Practice," *Transactions of the American Institute of Mining Engineers* XXIII, New York, 1894, 370; "Annual Report of the Secretary of Internal Affairs of the Commonwealth of Pennsylvania, Part III," *Industrial Statistics* XXII, no. 10, 1894, State Printer of Pennsylvania, 1895, D. 99.

136. Gayley, "The Development of American Blast Furnaces with Special Reference to Large Yields," 948–54.

137. Clark, *History of Manufactures in the United States*, 255.

138. Gayley, "The Development of American Blast Furnaces with Special Reference to Large Yields," 976.

139. Joseph Frazier Wall, *The Andrew Carnegie Reader* (Pittsburgh: University of Pittsburgh Press, 1992), 74.

140. Quote by Harry Huse Campbell, General Manager of the Pennsylvania Steel Company at Harrisburg and one of the chief competitors of the Carnegie Steel Company. Harry Huse Campbell, "The Manufacture and Properties of Iron and Steel," *Engineering and Mining Journal*, New York, 1903, 661.

141. Turner, *The Metallurgy of Iron and Steel*, 34.

142. "Manufacture of Pig Iron," *Annual Report of the Secretary of Internal Affairs of the Commonwealth of Pennsylvania*, D. 99–D. 102.

143. Jeans, *Reports of the Commissioners*, p. 22. Although the stated data is for foundry practice, this information can be related directly to the operation of cupolas for a converter house.

144. Henry M. Howe, "Notes on the Bessemer Process," 1150.

145. Robert W. Hunt, *A History of the Bessemer Manufacture in America*, 215.

146. Henry M. Howe, "Notes on the Bessemer Process"; Howe, *The Metallurgy of Steel*, 359.

147. Captain W. R. Jones, "On the Manufacture of Bessemer Steel and Steel Rails in the United States," 129.

148. Captain W. R. Jones, "On the Manufacture of Bessemer Steel and Steel Rails in the United States," 129–30.

149. *The Supreme Court Reporter* 22, West Publishing Co., St. Paul, 1902, 723.

150. Henry M. Howe, "Notes on the Bessemer Process," 1152.

151. Henry M. Howe, "Notes on the Bessemer Process."

152. *Carnegie Steel Co. v. Cambria Iron Co., The Federal Reporter*, October–December 1898, West Publishing Co., St. Paul, 1899, 756.

153. *Directory to the Iron and Steel Works of the United States* (Philadelphia: The American Iron and Steel Association, 1884), 115; Captain W. R. Jones, "On the Manufacture of Bessemer Steel and Steel Rails in the United States"; *Journal of the Association of Engineering Societies* I, 1882, 408.

154. Bridge, *The Inside History of the Carnegie Steel Company*, 69, 119–21, 135.

155. *Carnegie Steel Co. v. Cambria Iron Co.*, 756–57.

156. "Pig Iron Mixers," *Journal of the Iron and Steel Institute* LXII, E. & F. N. Spon, London, 1903, 475–76.

157. *The Iron Age*, May 22, 1902, 7.

158. Obituary of William Richard Jones, 180.

159. *The Iron Trade Review* XLIV, Penton Publishing Co., Cleveland, Ohio, 1909, 58.

160. *Railway World* 15, no. 40, Philadelphia, 1889, 939.

161. *American Manufacturer and Iron World* XL, May 13, 1887, 7.

162. *Pittsburgh Commercial Gazette*, "Grief at Braddock," September 30, 1889, 2.

163. *Pittsburgh Press*, "Capt. Jones Dead," September 29, 1889.

164. Casson, *The Romance of Steel*, 33; Moses King, *King's Handbook of the United States*, Moses King Corporation, Buffalo, New York, 1891–1892, 748–49; *Pittsburgh Commercial Gazette*; *Pittsburgh Press*; *American Manufacturer and Iron World* XLIV, no. 23, October 4, 1889, 7.

165. Bridge, *The Inside History of the Carnegie Steel Company*, 105; *Transactions of the American Society of Mechanical Engineers*, 841–42.

166. *Transactions of the American Society of Mechanical Engineers*, 842.

167. Henry Alford, *The Works of John Donne, D.D., Dean of Saint Paul's, 1621–1631*, Vol. III (London: John W. Parker, 1839), 575.

168. Frank Cowan, *Southwestern Pennsylvania in Song and Story* (Greensburg, PA: Printed by the Author, 1878), 38–39; Bridge, *The Inside History of the Carnegie Steel Company*, 85–86.

CHAPTER 6. PITTSBURGH BESSEMER STEEL COMPANY AND OPEN HEARTH STEELMAKING

1. James Howard Bridge, *The Inside History of the Carnegie Steel Company* (New York: The Aldine Book Company, 1903), 66–69.

2. William A. Berkey, *The Money Question* (Grand Rapids: W. W. Hart, Steam Book and Job Printer, 1876), 256–57; William Worthington Fowler, *Inside Life in Wall Street* (Hartford: Dustin Gilman & Company, 1873), 579–82.

3. Joseph Frazier Wall, *Andrew Carnegie* (New York: Oxford University Press, 1970).

4. Bridge, *The Inside History of the Carnegie Steel Company*, 85–86.

5. Bridge, *The Inside History of the Carnegie Steel Company*, 85–87 and 150–51.

6. Bridge, *The Inside History of the Carnegie Steel Company*, 150–51.

7. James M. Swank, *Introduction to a History of Ironmaking and Coal Mining in Pennsylvania* (Philadelphia: Published by the Author, 1878), 80.

8. *Annual Report of the Secretary of Internal Affairs of the Commonwealth of Pennsylvania, Part III Industrial Statistics* X, 1881–1882, Lane S. Hart State Printer and Binder, Harrisburg, Pennsylvania, 1883, Legislative Document No. 7, 34.

9. Swank, *Introduction to a History of Ironmaking and Coal Mining in Pennsylvania*, 83.

10. Bridge, *The Inside History of the Carnegie Steel Company*, 3.

11. Bridge, *The Inside History of the Carnegie Steel Company*, 3–5.

12. Bridge, *The Inside History of the Carnegie Steel Company*, 8, 16–19.

13. Bridge, *The Inside History of the Carnegie Steel Company*, 21–23.

14. Joseph Frazier Wall, *Andrew Carnegie* (New York: Oxford University Press, 1970), 340.

15. Bridge, *The Inside History of the Carnegie Steel Company*, 5–8.

16. Mark M. Brown, *The Architecture of Steel: Site Planning and Building Type in the Nineteenth-Century American Bessemer Steel Industry*, Unpublished Doctoral Thesis, University of Pittsburgh, 1995, 174–75.

17. Wall, *Andrew Carnegie*, 347.

18. Brown, *The Architecture of Steel*, 174–79.

19. James M. Swank, *Statistics of the American and Foreign Iron Trades, Annual Report of the Secretary of the American Iron and Steel Association*, American Iron and Steel Association, Philadelphia, 1881, 25; Kloman Eulogy, *Proceedings of the Engineers Society of Western Pennsylvania* I, Pittsburgh, 1880, 250–53; Memorial of Alexander Lyman Holley, C.E., LL.D., Born, July 20, 1832, Died January 29, 1882, American Institute of Mining Engineers, New York City, 1884.

20. Krause, Ibid, 172–73; Wall, *Andrew Carnegie*, 487.

21. George E. McNeill, ed., *The Labor Movement: The Problem of To-Day* (Boston: A. M. Bridgman & Co.; New York: M. W. Hazen Co., 1887), 268–71.

22. McNeill, *The Labor Movement*, 282–84.

23. *Eleventh Annual Report of the Bureau of Statistics of Labor to the General Assembly of the State of Ohio for the Year 1887* (Columbus, OH: Westbote Co., State Printers, 1888), 31.

24. McNeill, *The Labor Movement*, 263–64.

25. Krause, Ibid, 209–214.

26. John Percy, *Metallurgy: Iron and Steel* (London: John Murray, 1864), 794.

27. William Metcalf, “Revolution in Steel-Making,” *The Engineering Magazine* IX, April to September 1895, New York, 1075.

28. W. T. Jeans, *The Creators of the Age of Steel* (New York: Charles Scribner’s Sons, 1884), 131–33; George W. Maynard, *Biographical Notice of Sir William Siemens* (Privately Published, 1884), 1–9.

29. The Commissioners of Patents, Abridgements of the Specifications Relating to the Manufacture of Iron and Steel, The Great Seal Patent Office, London, 1858, 246.

30. Edward A. Cowper, “On Some Regenerative Hot-Blast Stoves Working at a Temperature of 1300° Fahrenheit,” *Newton’s Journal of Arts and Sciences* XII, London, 1860, 165–67; The Commissioners of Patents, 257–58.

31. Percy, *Metallurgy: Iron and Steel*, 674–75.

32. Percy, *Metallurgy: Iron and Steel*.

33. Percy, *Metallurgy: Iron and Steel*.

34. Samuel Kneeland, ed., *Annual of Scientific Discovery* (Boston: Gould and Lincoln, 1868), 14.

35. Percy, *Metallurgy: Iron and Steel*, 675; E. F. Bamber, ed., *The Scientific Works of C. William Siemens Kt.*, volume I (London: John Murray, 1889), 219.

36. The Commissioners of Patents.

37. W. T. Jeans, *The Creators of the Age of Steel*, 155.

38. W. T. Jeans, *The Creators of the Age of Steel*, 160.

39. William Henry Greenwood, *A Manual of Metallurgy* (London and Glasgow: William Collins, Sons, & Company, 1874), 154–55 and 127–31; J. T. Smith, "On Siemens Regenerative Furnace and Application to Re-Heating Furnaces," *Transactions, Iron and Steel Institute*, volume I, M. and M. W. Lambert, New Castle-Upon-Tyne, 1870, 145–48; Percy, *Metallurgy: Iron and Steel*, 673–79.

40. Percy, *Metallurgy: Iron and Steel*, 674.

41. How Products Are Made, Inventor Biographies, http://www.madehow.com/inventorbios/42/Pierre-Emile-Martin.html; Economypoint.org, Siemens-Martin open hearth furnace, History, http://www.economypoint.org/s/siemens-martin-open-hearth-furnace.html.

42. M. L. Gruner, *The Manufacture of Steel*, trans. Lenox Smith (New York: D. Van Nostrand, 1872), 82–83.

43. J. S. Jeans, *Steel: Its History, Manufacture, Properties and Uses* (London: E. & F. N. Spon, 1880), 92.

44. Swank, *Introduction to a History of Ironmaking and Coal Mining in Pennsylvania*, 85.

45. *American Annual Cyclopaedia and Register of Important Events of the Year 1868*, volume VIII (New York: D. Appleton and Company, 1869), 470.

46. Bamber, Ibid, p. 230.

47. W. T. Jeans, *The Creators of the Age of Steel*, 155.

48. W. T. Jeans, *The Creators of the Age of Steel*, 157.

49. H. O. Hofman, *An Outline of the Metallurgy of Iron and Steel* (Boston: Thomas Todd Printer (for MIT), 1904), 169–70.

50. There was no one formula or set of instructions for making steel. The example assembled was created from a mixture of the following sources: Greenwood, *A Manual of Metallurgy*, 178–80; W. T. Jeans, *The Creators of the Age of Steel*, 158–59; *The Journal of the Franklin Institute* LIV, Franklin Institute, Philadelphia, 1867, 257–58; William Crookes and Ernst Rohrig, *A Practical Treatise on Metallurgy*, volume III (London: Longmans, Green and Co., 1870), 207–8; H. Bauerman, *A Treatise on the Metallurgy of Iron* (New York: Virtue and Yorston, 1868), 401–3.

51. *First Helper's Manual*, Basic Incorporated, Dolomitic Refractories Association, Cleveland, Ohio, 1956, 31.

52. *The Journal of the Franklin Institute*.

53. *The Journal of the Iron and Steel Institute* No. 2, 1886, E. & F. N. Spon, London, 953.

54. J. M. Camp and C. B. Francis, *The Making, Shaping and Treating of Steel, Second Edition*, The Carnegie Steel Company, Pittsburgh, PA, 1920, p. 225..

55. Arthur H. Hiorns, *Iron and Steel Manufacture* (London: MacMillan and Co., 1889), 153–54; Abram S. Hewitt, *The Production of Iron and Steel in its Economic and Social Relations*, Paris International Exhibition, 1867, Government Printing Office, Washington, 1868, 25–26.

56. W. T. Jeans, *The Creators of the Age of Steel*, 158–59.

57. Greenwood, *A Manual of Metallurgy*, 178. Crookes and Rohrig, *A Practical Treatise on Metallurgy*, 203.

58. Greenwood, *A Manual of Metallurgy*, 180.

59. R. Howson, "On the Siemens-Martin Process of Manufacturing Cast Steel," *The Iron and Steel Institute Transactions* I, M. and M. W. Lambert, New Castle-Upon-Tyne, 1870, 177.

60. Thomas Turner, *The Metallurgy of Iron* (London; Charles Griffin & Company, Ltd., 1908), 47–48; J. M. Camp and C. B. Francis, *The Making, Shaping and Treating of Steel*, second edition (Pittsburgh: The Carnegie Steel Company, 1920), 177; K. C. Barraclough, *Steelmaking: 1850–1900* (London: The Institute of Metals, 1990), 206–13.

61. Barraclough, *Steelmaking: 1850–1900.*

62. M. Harmet, "On Dephoshorisation in the Bessemer Converter by Method of Repouring," *Journal of the Iron and Steel Institute*, E. & F. N. Spon, London, 1879, 155.

63. David Carnegie and Sidney C. Gladwyn, *Liquid Steel: Its Manufacture and Cost* (London: Longmans, Green and Co., 1913), 263.

64. *The Journal of the American Foundrymen's Association* 4, no. 19 (January 1898), Detroit, MI, 198.

65. William Serrin, *Homestead: The Glory and Tragedy of an American Steel Town* (New York: Times Books/Random House, 1992), 35–36.

66. Serrin, *Homestead.*

67. Kenneth Warren, *Industrial Genius* (Pittsburgh: University of Pittsburgh Press, 2007), 9.

68. Wall, *Andrew Carnegie*, 348.

69. *Directory to the Iron and Steel Works of the United States*, The American Iron and Steel Association, Philadelphia, 1888, 115.

70. *American Manufacturer and Iron World* XXXIX, July 16, 1886, Pittsburgh, 7; Bridge, *The Inside History of the Carnegie Steel Company*, 93.

71. Henry Marion Howe, "The Metallurgy of Steel," *The Engineering and Mining Journal*, New York, 1904, 331; *The Journal of the Iron and Steel Institute* XXX, E. & F. N. Spon, Ltd., London, 1887, 423.

72. *The Journal of the Iron and Steel Institute* XXX; W. Richards and J. A. Potter, "The Homestead Steel Works," *Proceedings of the United States Naval Institute* XV, no. 3, Annapolis, MD, 1889, 432–33.

73. Wall, *Andrew Carnegie*, 418–19.

74. Serrin, *Homestead.*

75. *American Manufacturer and Iron World* XXXVIII, April 23, 1886, Pittsburgh, 7.

76. Earnshaw Cook, *Open-Hearth Steel Making* (Cleveland: American Society for Metals, 1937), 28; William C. Buell Jr., *The Open Hearth Furnace*, volume I (Cleveland: The Penton Publishing Co., 1936), 5–6.

77. *American Manufacturer and Iron World*, April 23, 1886.

78. Cook, *Open-Hearth Steel Making.*

79. *The Journal of the Iron and Steel Institute in America in 1890*, E. & F. N. Spon, London, Special Volume of Proceedings, 1890, 276.

80. Stephen L. Goodale, *Chronology of Iron and Steel* (Pittsburgh: Pittsburgh Iron and Steel Foundries Company, 1920), 202.

81. George H. Thurston, *Allegheny County's Hundred Years* (Pittsburgh: A. A. Anderson & Son, 1888), 149.

82. George B. Hill, *Pittsburgh: Its Commerce and Industries, and the Natural Gas Interest* (Pittsburgh: J. Eichbaum, & Co., 1887), 15.

83. *Directory to the Iron and Steel Works of the United States* (Pittsburgh: The American Iron and Steel Association, 1886), 104–15.

84. Hill, *Pittsburgh: Its Commerce & Industries & the Natural Gas Interest*, 7–8.

85. *Iron Age* 61 (June 30, 1898): 12.

86. Park Benjamin, ed., *Modern Mechanism* (London: MacMillan and Co., 1892), 809.

87. Cook, *Open-Hearth Steel Making*.

88. James M. Swank, *History of the Manufacture of Iron in All Ages* (Philadelphia: The American Iron and Steel Association, 1892), 424; Leo F. Reinartz, *Steelmaking/USA: History of Iron and Steelmaking in the United States* (New York: The Metallurgical Society, American Institute of Mining Metallurgical and Petroleum Engineers, New York, 1961), 76, publication in one book of a series of historical articles that have appeared in *Journal of Metals*, 1956–1961; Goodale, *Chronology of Iron and Steel*, 203–4.

89. Barraclough, *Steelmaking: 1850–1900*.

90. T. Egleston, "Basic Open Hearth Steel Process," *School of Mines Quarterly* VII, Columbia College, New York, 1886, 49.

91. *The Journal of the Iron and Steel Institute* LII, no. 1, E. & F. N. Spon, Ltd., London, 1898, 489–90; Reinartz, *Steelmaking/USA*.

92. Cook, *Open-Hearth Steel Making*, 27; S. T. Wellman, "The Early History of Open Hearth Steel Manufacture in the United States," *Transaction of the American Society of Mechanical Engineers* XXIII, American Society of Mechanical Engineers, New York, 1902, 97.

93. *Directory to the Iron and Steel Works of the United States*, xi and 19.

94. Richards and Potter, "The Homestead Steel Works," 437.

95. *American Manufacturer and Iron World* XXXIX, October 29, 1886, Pittsburgh, 7.

96. *American Manufacturer and Iron World* XXXIX, December 31, 1886, Pittsburgh, 9.

97. *Railway World* 14, Philadelphia, 1888, 536.

98. *Official Gazette of the United States Patent Office* XL, July 5 to September 27, Inclusive, 1887, Government Printing Office, Washington, 1888, 50.

99. Charles Reitell, *Machinery and Its Benefits to Labor in the Crude Iron and Steel Industries*, Doctoral Thesis, University of Pennsylvania, The Collegiate Press, George Banta Publishing Company, Menasha Wisconsin, 1917, 21; *Proceedings of the United States Naval Institute* XV, no. 3, 437.

100. Harry Huse Campbell, *The Manufacture and Properties of Structural Steel* (New York: The Scientific Publishing Co., 1896), 97.

101. L. W. (Anonymous), "Homestead as Seen by One of Its Workmen," *McClure's Magazine* III, 1894, S. S. McClure Limited, New York and London, 167.

102. Reitell, *Machinery and Its Benefits to Labor*, 23–24.

103. Reitell, *Machinery and Its Benefits to Labor*.

104. Alfred E. Hunt, "Some Recent Improvements in Open-Hearth Practice," *Transactions of the American Institute of Mining Engineers* XVI, American Institute of Mining Engineers, New York, 1888, 702.

105. L. W., "Homestead as Seen by One of Its Workmen," 168.

106. Reitell, *Machinery and Its Benefits to Labor*, 24.

107. L. W., "Homestead as Seen by One of Its Workmen."

108. *The Engineering and Mining Journal* XLIX, January to June 1890, Scientific Publishing Company, New York, 214; *The Iron and Steel Institute in America in 1890 (Special Volume of the Proceedings)*, E. & F. N. Spon, London, 1890, 434.

109. *American Economist (The Tariff Review)* X, July 15, 1892, The American Protective Tariff League, New York, 38–39; *The Iron and Steel Institute in America in 1890 (Special Volume of the Proceedings)*.

110. *The Journal of the Iron and Steel Institute in America in 1890*, 434.

111. *Transactions* XIX, American Institute of Mining Engineers, New York, 1891, 316–17; *Journal of the Iron and Steel Institute* 37, E. & F. N. Spon, London, 1890, 784–85.

112. *Scientific American Supplement* XLIII, no. 1119, June 19, 1897, New York, 17890; *Engineering* LXIII, January–June 1897, London, 658–59; *Transactions of the American Institute of Mining Engineers* XIX, American Institute of Mining Engineers, New York, 1891, 316–17; Jeremiah Head, "On Charging Open Hearth Furnaces by Machinery," *The Electrical Engineer* XX, London, 1897, 409.

113. Photographic evidence showing the form of construction of these cars and charging boxes used at Carnegie's Homestead Works dating from the period of 1893 to 1895 can be found in the collection of the Carnegie Museum of Art, photo identifier number 1999.19.3.

114. Ernest Kilburn Scott, *The Local Distribution of Electric Power in Workshops* (London: Biggs and Co., 1897), 130–32; *Engineering* LXIII, 669. (Mr. Windsor-Richards stated that he had seen the machines in operation at Homestead a few years before the publication of the paper in 1897.)

115. *Scientific American Supplement* XLIII, no. 1119, 17891.

116. *Engineering* LXIII, 658–59.

117. *Minutes of the Proceedings of the Institution of Civil Engineers* CXXXIV, London, 1898, 461.

118. *The Railway and Engineering Review* XXXVII, 1897, Chicago, 88.

119. Catalog, The Wellman-Seaver Engineering Co. Engineers and Manufacturers, Cleveland, Ohio, 1901, 43–45.

120. *Engineering* LXIII, 658 and 660.

121. Jeremiah Head, "On Charging Open Hearth Furnaces by Machinery," 406.

122. Determined from photographic evidence of two electrically driven overhead traveling cranes used at Carnegie's Homestead Works dating from the period of 1893 to 1895 found in the collection of the Carnegie Museum of Art, photo identifier numbers 1999.19.13 and 1999.19.27. The crane in photo 1999.19.13 is identified as one built by the Morgan Engineering Company of Alliance, Ohio, with a construction number of 630 and, according to a purchase order in Morgan's files dated January 21, 1893, it was converted by them to operate electrically. Prior to that date, the records

show the existing crane was belt and pulley driven. This information was determined through personal communication with Morgan's Engineering Department. The other crane, 1999.19.27, was one built by William Sellers Company of Philadelphia for the armor plate casting pit (no construction date information was found) in the #2 open hearth shop and had a capacity of sixty-none tons.

123. Richards and Potter, "The Homestead Steel Works," 439.

124. The Metallurgical Society, *History of Iron and Steelmaking in the United States* (publication in one book of a series of historical articles that have appeared in *Journal of Metals*, 1956–1961) (New York: American Institute of Mining, Metallurgical and Petroleum Engineers), 80.

125. Christopher J. Dawson, *Steel Remembered: Photographs From the LTV Steel Collection* (Cleveland: Kent State University Press, 2008), 137. A photo from the Western Reserve Historical Society's LTV Steel Collection, Republic negative number C23, shows a hydraulic crane teeming ingots with a Bessemer converter in operation in the background at the Republic Steel Company, Youngstown, Ohio, plant. The photo was dated circa 1940s.

126. *Western Electrician* XXI, August 28, 1897, Electrician Publishing Company, Chicago, 117.

127. *The Engineering Magazine* XXIV, Engineering Magazine Co., New York, 1903, 402–15.

128. BLH Dabbs photo collection Carnegie Museum of Art (COMA), Pittsburgh.

129. Charles Scott, "A Half Century of Electrical Engineering," *Proceedings of the Engineers' Society of Western Pennsylvania* 47, Engineers' Society of Western Pennsylvania, Pittsburgh, 1931, 227; D. Selby-Bigge, "Electricity in the Iron and Steel Industries," *Engineering Magazine* VIII, Engineering Magazine Co., New York, 1895, 407; Richards and A. Potter, "The Homestead Steel Works," 436.

130. Richards and Potter, "The Homestead Steel Works," 441.

131. F. Ernest Johnson, "The Twelve Hour Day in the Steel Industry," pamphlet, Federal Council of the Churches of Christ in America, New York, 1923, 18.

132. *The Manufacturer and Builder* XXVI, April 1892, New York, 96.

133. *Iron Age* 61, June 30, 1898, 12–14. Photos of the plant are in a supplement.

134. D. R. Loughrey, *Scrap Charging in Steelmaking Furnaces, Iron and Steel Engineer Yearbook 1962*, Association of Iron and Steel Engineers Yearly Proceedings, AISE, Pittsburgh, 1962, 326.

135. Bridge, *The Inside History of the Carnegie Steel Company*, 98.

136. Cook, *Open-Hearth Steel Making*.

137. Campbell, *The Manufacture and Properties of Structural Steel*, 156.

138. *Second Report of the Bureau of Mines 1892*, Sessional Papers No. 35, Legislative Assembly of Ontario, Toronto, 1893, 56.

139. "Greater Pittsburgh and Allegheny County," *American Manufacturer and Iron World*, 1901. Published for distribution at the Pan-American Exposition in Buffalo, New York, in 1901. According to the section subtitled "The Huge Enterprises Built Up by Andrew Carnegie," it indicated that the Homestead Works open hearth output was 34 percent of that of the United States in 1900. No page numbers were included, but including cover, page 47. *Report of the Industrial Commission of Trusts*

and Industrial Combinations, second volume), Volume XIII of Commission's Reports, Government Printing Office, Washington, DC, 1901, XCVII; *The Iron Age* 61, June 30, 1898, David Williams, New York, 12.

140. Alfred Davis, "The Consolidation of Fluid Steel," *Van Nostrand's Eclectic Engineering Magazine* XXI, D. Van Nostrand, New York, 1879, 469–70.

141. *First Helper's Manual*, 33.

142. Prices of Armor for Navy Vessels, Government Printing Office, Washington, 1896, 189.

143. Frank Popplewell, *Some Modern Conditions and Recent Developments in Iron and Steel Production in America* (Manchester, UK: Manchester University Press, 1906), 101.

144. *History of Iron and Steelmaking in the United States*, Metallurgical Society of the AIME, 78. This is a publication in one book of a series of articles found in the *Journal of Metals* between 1956 and 1961. The section cited is by Leo F. Reinartz.

145. J. G. Eaton, "Domestic Steels for Naval Purposes," *Proceedings of the United States Naval Institute* XV, Annapolis, MD, 1889, 319.

CHAPTER 7. DUQUESNE

1. Joseph Frazier Wall, *Andrew Carnegie* (New York: Oxford University Press, 1970), 497–98; James Howard Bridge, *The Inside History of the Carnegie Steel Company* (New York: The Aldine Book Company, 1903), 174–80.

2. Wall, *Andrew Carnegie*, 499.

3. "Blast Furnace Practice," *The New American Supplement Encyclopedia Britannica*, Vol. III (New York: The Werner Company, 1897), 1696.

4. J. E. Johnson Jr., *Blast Furnace Construction in America* (New York: McGraw Hill Book Company, 1917), 15.

5. *The Engineer* LXXIII, January to June 1892, London, 213; *Journal of the Iron and Steel Institute*, no. 1, E. & F. N. Spon, London, 1892, 400.

6. J. E. Johnson Jr., *Filling the Blast Furnace, Mechanical & Chemical Engineering*, Vol. XIII (New York: McGraw Publishing Company Inc., 1915), 165.

7. Johnson, *Blast Furnace Construction in America*, 15–16.

8. *Engineering* LXV, London, 1898, 381.

9. *Railway and Engineering Review* XXXVII, Chicago, 1897, 191.

10. *Engineering* LXIII, London, 1897, 472.

11. Johnson, *Filling the Blast Furnace*, 227.

12. *Engineering* LXIII; *Engineering* LXV, 1898, 382; Archibald P. Head, *Notes on American Iron and Steel Practice* (Brierley-Hill: Ford and Addison, 1898), 8.

13. *The Official Gazette of the United States Patent Office* 78, no. 9, March 28, 1897, 1694–95. Patent number 579,011, items eight through sixteen, for a description of positioning and lowering the bucket above the bell.

14. J. Stephen Jeans, ed., *American Industrial Conditions and Competition*, British Iron Trade Association, London, 1902, 473; H. O. Hofman, *An Outline of the*

Metallurgy of Iron and Steel (Boston: Thomas Todd Printer (for MIT), 1904), 49–50; Johnson, *Filling the Blast Furnace*.

15. *Engineering* LXIII.

16. *Engineering* LXV; *Engineering* LXIII, 538.

17. *Engineering* LXIII.

18. Jeans, *American Industrial Conditions and Competition*, 473. This report states that the inclined skip hoist was installed at Lucy in 1879 with a double skip. Several other sources (referenced elsewhere) indicate that an automatic system with skip cars was put into operation in 1884.

19. *Engineering and Mining Journal* XXXVII, Scientific Publishing Company, New York, 1884, 421; *American Machinist* 7, no. 24, June 14, 1884, American Machinist Publishing Company, New York, 5; *Engineering and Mining Journal* XXXVIII, Scientific Publishing Company, New York, 1884, 130.

20. Jeans, *American Industrial Conditions and Competition*.

21. F. Louis Grammer, *A Decade of American Blast Furnace Practice, Transactions of the American Institute of Mining Engineers*, vol. XXXV (New York: American Institute of Mining Engineers, 1905), 128–29.

22. *The Official Gazette of the United States Patent Office*; Jeans, *American Industrial Conditions and Competition*; *Transactions of the American Institute of Mining Engineers* XXVIII, New York, 1899, xxv and xxviii.

23. Johnson, *Filling the Blast Furnace*, 169.

24. Johnson, *Filling the Blast Furnace*, 227.

25. Johnson, *Filling the Blast Furnace*.

26. Johnson, *Filling the Blast Furnace*, 169; Jeans, *American Industrial Conditions and Competition*.

27. Johnson, *Filling the Blast Furnace*, 166.

28. David Parker, "Improvements in the Mechanical Charging of the Modern Blast-Furnace," *Transactions of the American Institute of Mining Engineers* XXXV, New York, American Institute of Mining Engineers, 1905, 554.

29. Johnson, *Filling the Blast Furnace*, 165.

30. Parker, "Improvements in the Mechanical Charging."

31. Charles Reitell, *Machinery and It's Benefits to Labor in the Crude Iron and Steel Industries*, Doctoral Thesis, University of Pennsylvania, The Collegiate Press, George Banta Publishing Company, Menasha, Wisconsin, 1917, 20.

32. *The Locomotive* XVI, Hartford Steam Boiler Insurance and Inspection Co., Hartford, Connecticut, 1895, 138–39.

33. Parker, "Improvements in the Mechanical Charging," 555; Grammer, *A Decade of American Blast Furnace Practice*, 134; H. O. Hofman, *An Outline of the Metallurgy of Iron and Steel*, 49; Johnson, *Filling the Blast Furnace*.

34. Johnson, *Filling the Blast Furnace*, 167 and 227.

35. Parker, "Improvements in the Mechanical Charging," 566–69; David Baker, "Stock Distribution and Its Relation to the Life of a Blast Furnace Lining," *Transactions of the American Institute of Mining Engineers* XXXV, New York, American Institute of Mining Engineers, 1905, 248–55; Johnson, *Filling the Blast Furnace*, 167.

36. Johnson, *Filling the Blast Furnace*, 172; Parker, "Improvements in the Mechanical Charging," 563–65; Baker, "Stock Distribution and Its Relation to the Life of a Blast Furnace Lining."

37. Johnson, *Filling the Blast Furnace*, 167.

38. Carl A. Meissner, *The Modern By-Product Coke Oven*, United States Congressional Serial Set No. 6536, Senate Documents Vol. 21, Document No. 145, Government Printing Office, Washington, 1913, 21–22.

39. Johnson, *Filling the Blast Furnace*, 169–70; Charles H. Wright, "Ore Handling Plant at the Clairton Works of the Crucible Steel Company," *Journal of the Association of Engineering Societies* XXXIII, Association of Engineering Societies, Philadelphia, 1904, 26–27.

40. David Baker, "Improvements in the Mechanical Charging of the Modern Blast Furnace," *Transactions of the American Institute of Mining Engineers* XXXV, American Institute of Mining Engineers, New York, 1905, 565.

41. Johnson, *Filling the Blast Furnace*, 167; Baker, "Stock Distribution and Its Relation to the Life of a Blast Furnace Lining," 245–46.

42. Johnson, *Filling the Blast Furnace*, 227.

43. Johnson, *Filling the Blast Furnace*, 166.

44. "Edgar Thomson Steel Works: Mother Steel Plant of Carnegie Steel," *U.S. Steel News* 1, no. 4, United States Steel Corporation and Subsidiaries, Hoboken, New Jersey, 1936, 18.

45. *Iron and Steel Magazine* IX, Albert Sauveur, ed., Boston, 1905, 466.

46. "Greater Pittsburgh and Allegheny County," *American Manufacturer and Iron World*, 1901. Published for distribution at the Pan-American Exposition in Buffalo, New York, in 1901, this pamphlet contained a short biography of Julian Kennedy. No pagination.

47. James Howard Bridge, *The Inside History of the Carnegie Steel Company* (New York: The Aldine Book Company, 1903), 69–70, 93, 109, 161–62.

48. *A History of Cleveland Ohio*, Vol. III (Chicago-Cleveland: S. J. Clarke Publishing Co., 1910), 453–54.

49. Johnson, *Filling the Blast Furnace*, 172.

50. Julian Kennedy, "Some Modifications in Blast Furnace Construction," *Proceedings of the Engineers' Society of Western Pennsylvania* XXIII, Engineers' Society of Western Pennsylvania, Pittsburgh, Pennsylvania, 1908, 3–5.

51. Kennedy, *Some Modifications in Blast Furnace Construction*, 5.

52. Axel Sahlin, "A New Blast Furnace Top," *Engineering* LXV, London, 1903, 732.

53. Kennedy, *Some Modifications in Blast Furnace Construction*, 5–7; Sahlin, "A New Blast Furnace Top."

54. Kennedy, *Some Modifications in Blast Furnace Construction*, 12; Sahlin, "A New Blast Furnace Top," 731–32.

55. Kennedy, *Some Modifications in Blast Furnace Construction*, 5; Sahlin, "A New Blast Furnace Top," 732; Julian Kennedy, *Fifty Years of Mechanical Engineering, Proceedings of the Engineers' Society of Western Pennsylvania* 47, Engineers' Society of Western Pennsylvania, Pittsburgh, 1931, 211–13.

56. Sahlin, "A New Blast Furnace Top."

57. *Engineering* LXIII, 472; *Engineering News* XXXVII, Engineering News Publishing Co., New York, 1897, 270.
58. E. A. Uehling, "Pig Iron Casting and Conveying Machinery: Its Development in the United States," *Cassier's Magazine* XXIV, The Cassier Magazine Company, New York, 1903, 115–17.
59. Uehling, "Pig Iron Casting and Conveying Machinery," 117–18.
60. Uehling, "Pig Iron Casting and Conveying Machinery," 119–22.
61. Uehling, "Pig Iron Casting and Conveying Machinery," 123.
62. Uehling, "Pig Iron Casting and Conveying Machinery."
63. Uehling, "Pig Iron Casting and Conveying Machinery."
64. *Engineering* LXIII, 473; *The Railway and Engineering Review*, 192.
65. Charles Scott, "A Half Century of Electrical Engineering, Proceedings of Engineer's Society of Western Pennsylvania," *Engineer's Society of Western Pennsylvania* 47 Pittsburgh, 1931, 225.
66. Ernest Kilburn Scott, *The Local Distribution of Electric Power in Workshops* (London: Biggs and Co., 1897), 126.
67. E. A. Holbrook, "Progress in Coal Mining in the Pittsburgh District," *Proceedings of Engineer's Society of Western Pennsylvania* 47, Engineer's Society of Western Pennsylvania, Pittsburgh, 1931, 222.
68. *Journal of the Iron and Steel Institute* LXII, 1903, 514.
69. *Iron Age* LXXI, January 1, 1903, 12–14.
70. *Iron and Machinery World (The Age of Steel)* 89, no. 2, January–June 1901, St. Louis, 17.
71. Arthur C. Johnson, "American Ore Dock Machinery," *Cassier's Magazine* XVIII, no. 5, September 1900, 369; *Iron and Machinery World*, 17–18; *Scientific American* LXXXIV, March 9, 1901, 153; *Marine Review* XIX, no. 12, March 28, 1899, 15; *Marine Review* XX, no. 22, November 30, 1899, 20.
72. Arthur C. Johnson, "American Ore Dock Machinery," 366.
73. Arthur C. Johnson, "American Ore Dock Machinery," 361; *Iron and Machinery World*, 17; *The Iron Age*, April 5, 1900, 3.
74. *The Iron Age*, April 5, 1900, 1–2.
75. *Scientific American*, LXXXIV.
76. Arthur C. Johnson, "American Ore Dock Machinery," 351; *The Iron Age*, April 5, 1900, 3.
77. *Iron and Machinery World*, 18.
78. *Iron and Machinery World*, 17.
79. Arthur C. Johnson, "American Ore Dock Machinery," 368–69.
80. *Marine Review* XX.
81. *Marine Review* XX.

CHAPTER 8. SUMMATION

1. James M. Swank, "The Manufacture of Iron and Steel Rails in Western Pennsylvania," *Pennsylvania Magazine of History and Biography* XXVII, no. 109 (January 1904), The Historical Society of Pennsylvania, Philadelphia, 3.

2. Charles R. Morris, *The Tycoons* (New York: Times Books, 2005), 319–30. Private communication with Kenneth Warren, January 25, 2013.

3. J. Stephen Jeans, "Future Supremacy in the Iron Markets of the World," *Engineering Magazine* XIV, New York, 1898, 199; *Proceeding of the Institution of Civil Engineers* CXXXIV, Institution of Civil Engineers, London, 1898, 461.

4. *Iron Age* LXVII, David Williams Company, New York, November 21, 1901, 52.

5. *The Michigan Miner* 2, no. 4, Saginaw, Michigan, March 1, 1900, 25.

6. Engineers' Society of Western Pennsylvania, *Transactions* I, Pittsburgh, 1882, 252.

7. "The Ironworks of the United States," American Iron and Steel Association, Philadelphia, 1876, 118.

8. *Railway World* 5, Philadelphia, 1879, 535; *Journal of the Iron and Steel Institute* 18, E. & F. N. Spon, London, 1881, 131–34; *Journal of the Iron and Steel Institute* 19, 370–71.

9. *Transactions of the American Institute of Mining Engineers* V, Easton, Pennsylvania, 1877, 209; J. S. Jeans, *Steel: Its History, Manufacture, Properties and Uses* E. & F. N. Spon, London, 1880, 146–48; *Directory to the Iron and Steel Works of the United States*, The American Iron and Steel Association, Philadelphia, 1878, 134.

10. Joseph Frazier Wall, *Andrew Carnegie* (New York: Oxford University Press, 1970), 521–26.

11. A. L. Holley and Lenox Smith, "American Steel Works," *Engineering* XXV, London, 1878, 295.

12. *Journal of the Iron and Steel Institute* 23, 1883, 747.

13. *Engineering* XXVIII, London, 1879, 84; *Journal of the Iron and Steel Institute* 15, 1879, 476–78.

14. *American Manufacturer and Iron World* 69, Pittsburgh, Pennsylvania, 1901, 1352.

15. *Iron Age* LXXIII, April 14, 1904, 18.

16. Swank, "The Manufacture of Iron and Steel Rails in Western Pennsylvania," 7–9.

17. *Directory of Iron & Steel Works of the United States*, eighth edition (Philadelphia: The American Iron and Steel Association, 1886), 178–80.

18. J. S. Jeans, "Future Supremacy in the Iron Markets of the World."

19. *Journal of the Iron and Steel Institute* LXII.

20. *Journal of the Iron and Steel Institute* 31, 1887, 314.

21. *Engineering* LXXIII, London, 1902, 173.

22. "Statistics of the American and Foreign Trades for 1901," American Iron and Steel Association, Philadelphia, 1902, 39–40; *Monthly Bulletin of the Bureau of the American Republics*, January 1902, Government Printing Office, Washington, DC, 1902, 1235.

23. *The Iron Trade Review* XLVII, Penton Publishing Company, Cleveland, OH, 1911, 627.

24. *American Manufacturer and Iron World*, May 10, 1889, Pittsburgh, Pennsylvania, 11.

25. James M. Swank, *Directory of Iron and Steel Works of the United States*, fifteenth edition (Philadelphia: American Iron and Steel Association, 1901), 101.

26. Harry Huse Campbell, "The Manufacture and Properties of Iron and Steel," *Engineering and Mining Journal*, New York, 1903, 205–13.

27. Edward Sherwood Meade, *Trust Finance* (New York and London: D. Appleton and Company, 1910), 210.

28. "Annual Report of the Secretary of Internal Affairs of the Commonwealth of Pennsylvania, Part III," *Industrial Statistics* VII, 1879–1880, Harrisburg, Pennsylvania, 1881, 41.

29. Jeans, "Future Supremacy in the Iron Markets of the World," 520.

30. Jeans, "Future Supremacy in the Iron Markets of the World," 542; *Journal of the Iron and Steel Institute* 40, 1892, 434.

31. *American Manufacturer and Iron World* XXXIX, July 16, 1886, Pittsburgh, 7.

32. Frederick T. Gretton Photographs, 1857–1953, MSP 328, Library and Archives Division, Senator John Heinz History Center, photo identifiers: MSP328.B001.F13.I02, MSP328.B001.F13.I04, MSP328.B001.F16.I01, MSP328.B001.F14.I06.

33. John N. Ingham, *Making Iron and Steel: Independent Mills in Pittsburgh, 1820–1920* (Columbus, Ohio: Ohio State University Press, 1991), 92–93.

34. Benjamin Talbot, "The Open Hearth Continuous Process," *Journal of the Iron and Steel Institute* LVII, E. & F. N. Spon, London, 1900, 33–44; Jeans, "Future Supremacy in the Iron Markets of the World," 534.

35. "History of Iron and Steelmaking in the United States," *Metallurgical Society of the AIME*, 79. This is a publication in one book of a series of articles found in the *Journal of Metals* between 1956 and 1961. The section cited is by Leo F. Reinartz.

36. Meade, *Trust Finance*, 210–11.

37. Kenneth Warren, *Industrial Genius* (Pittsburgh: University of Pittsburgh Press, 2007), 10–19.

38. Wall, *Andrew Carnegie*, 491–96 and 747–57.

39. *Timely Topics* IV, no. 25, February 23, 1900, 393.

40. Talbot, "The Open Hearth Continuous Process," 72.

41. Jeremiah Head, "On Charging Open Hearth Furnaces by Machinery," *Electrical Engineer* XX, London, 1897, 409.

42. Talbot, "The Open Hearth Continuous Process," 33; International Library of Technology, *Manufacture of Steel* (Section 11), International Textbook Company, Scranton, 1902, 65; David Carnegie, *Liquid Steel: Its Manufacture and Cost* (London: Longmans, Green, and Co., 1913), 264–65.

43. Talbot, "The Open Hearth Continuous Process," 40, 44, 73.

44. Talbot, "The Open Hearth Continuous Process," 75–79.

45. John Hays Smith, "Electricity in Modern Steel Making," *Engineering Magazine* XXIV, New York, 1903, 402–5; Malcolm MacLaren, *The Rise of the Electrical Industry during the Nineteenth Century* (Princeton: Princeton University Press, 1943), 96–97.

46. *The Bulletin* XXXV, American Iron and Steel Association, March 10, 1901, 37.

47. *Engineering and Mining Journal* LXX, Scientific Publishing Company, New York, 1900, 547.

48. David Brody, *Steelworkers in America: The Nonunion Era* (New York: Harper & Row, 1969), 6.

49. Brody, *Steelworkers in America,* 2.

50. Private communication with Kenneth Warren, January 13, 2013.

51. *Report of the Industrial Commission of Trusts and Industrial Combinations* (Second Volume), Volume XIII of Commission's Reports, Government Printing Office, Washington, DC, 1901, XCVII.

52. Wall, *Andrew Carnegie*, 385–86.

53. Wall, *Andrew Carnegie*, 104–5, 128.

54. Wall, *Andrew Carnegie*, 224–25.

55. Wall, *Andrew Carnegie*, 789–90.

Index

About the Author

Ken Kobus is a third-generation steelworker who retired after nearly forty-five years of service in the steel industry. His father (1937), paternal grandfather (1906), and he (1966) all worked for the Jones and Laughlin Steel Company on Pittsburgh's South Side. Together they achieved well over one hundred years of service at J&L and its successor company, LTV. His maternal grandfather also was also a steel man, starting in 1906 at the Carnegie Steel Company's Duquesne Works in the #1 open hearth shop. All of his grandparents were immigrants from Eastern Europe.

Ken began his career as a laborer and worked his way through various stages of management at the facility's metallurgical coke plant, ending as general foreman of the by-product (coal chemicals) plant before its closure. He was transferred to LTV's South Chicago coke plant as area manager of coking and coal handling. When the company declared bankruptcy, he returned to Pittsburgh to work for U.S. Steel at their Clairton Coke Plant, where he retired as the process safety manager. He holds a bachelor's degree in mechanical engineering from the University of Pittsburgh.

Resulting from a lifelong interest in trains, Ken is coauthor of three books and has authored a number of articles about the Pennsylvania Railroad in Pittsburgh. He also published an annotated version of a 1922 book titled *Steel: The Diary of a Furnace Worker*, where he established that it was written about operations at the Aliquippa Works of J&L. His interest in steel and steel history dates back to childhood. He convinced his father to take him to work, where he was permitted to tap his first open hearth furnace at age sixteen when they used explosives to open the tap hole. Ken also feels privileged to have personally witnessed Bessemer furnaces blowing at the National Tube Works (USS) in McKeesport, Pennsylvania, as well as beehive coke being made in the Connellsville coke district.

He is also an avid amateur astronomer. He resides in suburban Pittsburgh.